Disclaimer

The publisher of this book is by no way associated with the National Institute of Standards and Technology (NIST). The NIST did not publish this book. It was published by 50 page publications under the public domain license.

50 Page Publications.

Book Title: Performance Criteria For an ASTM XRF Standard Test Method For Chemical Analysis of Hydraulic Cements: Inter-Laboratory Study Cements E and F

Book Author: Paul E. Stutzman;

Book Abstract: Bulk oxide determinations from a pair of portland cements provides the basis for calculation precision and accuracy values for X-ray fluorescence (XRF) analysis for both the fused glass bead and the pressed powder sample preparation. This report is the second in a series on an Interlaboratory study on chemical analyses of hydraulic cements by X-ray fluorescence for the purpose of estimating precision and qualification criteria. Approximately 45 laboratories provided six replicates analyzed in duplicate for two separate portland cements containing ca. 5 % limestone, covering fifteen analytes, CaO, SiO2, Al2O3, Fe2O3, SO3, MgO, Na2O, K2O, TiO2, P2O5, Mn2O3, SrO, ZnO, Cr2O3, and Cl, with the laboratories roughly split between the two different sample preparations. Chemical data using traditional chemical analyses (the Reference Methods) from the Cement and Concrete Reference Laboratory (CCRL) proficiency test program were included for comparison to the XRF results. Precision measures for within- and between-laboratory performance are presented as 1σ and 95 % limits (ASTM d2s). Accuracy criteria are based upon a two-sided 95 % prediction interval for the mean of two test results, defining the range of values one might expect for each analyte relative to a certified value of a reference material.

Citation: NIST TN - 1815

Keywords: accuracy; hydraulic cement; precision; qualification; X-ray fluorescence

NIST Technical Note 1815

PERFORMANCE CRITERIA FOR AN ASTM XRF STANDARD TEST METHOD FOR CHEMICAL ANALYSIS OF HYDRAULIC CEMENTS: INTER-LABORATORY STUDY CEMENTS E AND F

Paul Stutzman
Alan Heckert

National Institute of
Standards and Technology
U.S. Department of Commerce

NIST Technical Note 1815

PERFORMANCE CRITERIA FOR AN ASTM XRF STANDARD TEST METHOD FOR CHEMICAL ANALYSIS OF HYDRAULIC CEMENTS: INTER-LABORATORY STUDY CEMENTS E AND F

Paul Stutzman
Materials and Structural Systems Division
Engineering Laboratory

Alan Heckert
Statistical Engineering Division
Information Technology Laboratory

October 2013

U.S. Department of Commerce
Penny Pritzker, Secretary

National Institute of Standards and Technology
Patrick D. Gallagher, Under Secretary of Commerce for Standards and Technology and Director

National Institute of Standards and Technology Technical Note 1815
Natl. Inst. Stand. Technol. Tech. Note 1815, 154 pages, October 2013
CODEN: NTNOEF

Abstract

Bulk oxide determinations from a pair of portland cements are used to calculate precision and accuracy values for X-ray fluorescence (XRF) analysis of both the fused glass bead and the pressed powder sample preparation methods. This report is the second in a series on an Interlaboratory study on chemical analyses of hydraulic cements by X-ray fluorescence for the purpose of estimating precision and qualification criteria. Approximately 45 laboratories provided six replicates analyzed in duplicate for two separate portland cements containing ca. 5 % limestone, covering fifteen analytes, CaO, SiO_2, Al_2O_3, Fe_2O_3, SO_3, MgO, Na_2O, K_2O, TiO_2, P_2O_5, Mn_2O_3, SrO, ZnO, Cr_2O_3, and Cl, with the laboratories roughly split between the two different sample preparations. Chemical data using traditional chemical analyses (the Reference Methods) from the Cement and Concrete Reference Laboratory (CCRL) proficiency test program were included for comparison to the XRF results. Precision measures for within- and between-laboratory performance are presented as 1σ and 95 % limits (ASTM d2s). Accuracy criteria are based upon a two-sided 95 % prediction interval for the mean of two test results, defining the range of values one might expect for each analyte relative to a certified value of a reference material.

Table of Contents

List of Tables

List of Figures

Introduction

Chemical analysis of portland cement is used for process control in clinker and cement manufacture, for demonstration specification compliance, and for relating chemical properties to performance attributes. Reporting requirements for the chemical composition of portland cements in North America date back to the *1915 Joint Conference on Uniform Methods of Tests and Standard Specifications for Cement,* which was written by the American Society of Civil Engineers, the American Society for Testing Materials (ASTM), and the United States Government, and included the following analytes: SiO_2, Fe_2O_3, Al_2O_3, CaO, MgO, SO_3, Ignition Loss, and Insoluble Residue [1].

In 1946, work on harmonizing the Federal and ASTM Cement Specifications resulted in the development of ASTM C150, Specification for Portland Cement within ASTM C1 on Cement. Standard Methods of Chemical Analysis of Portland Cement, ASTM C114-44 was also published and contained a performance-based qualification scheme [Table 1] that is very similar to that used today where duplicate determinations on different days are made and the two results must be within the limit of permissive variation with their average accepted as the correct value [2]. Reference cements were used to qualify each laboratory's process, and the standard method required the laboratory to repeat the measurement process if the qualification criteria were not met.

The 1960's through the 1970's saw a shift from classical analytical "wet" chemistry measurements (referred to as the 'reference-wet' and the 'alternate-wet' methods) to instrumental methods, including atomic absorption spectrometry, X-ray spectrometry, and a spectrophotometric/titrimetric scheme. Forrester et al. [3], Midgley [4], Harrison et al. [5], Aldridge et al. [6,7], Stutzman and Lane [8], European standard EN 196-2.2 [9], and a National Cooperative Highway Research Program report [10] have investigated uncertainty in chemical analyses of portland cements, with the latter three being based upon standardized methods. ASTM Subcommittee C01.23, Chemical Analysis, amended the precision requirements with accuracy requirements in 1977 by replacing the third column in Figure 1 with a maximum difference between the mean of two replicate determinations and the value of a certified reference material. This amended set of criteria is now Table 1 of ASTM C114 and is the basis of method qualification for instrumental methods such as, for example, those by X-ray fluorescence (XRF) [2]. While the column one requirements originated in the 1946 edition of ASTM C114 and are based upon the reference methods, none of the qualification requirements have supporting data available. The development of a standard test for XRF analysis of hydraulic cements provides an opportunity to collect the data and calculate appropriate qualification values.

Table 1 ASTM C114 performance criteria for the chemical analysis of portland cements from the Report of Working Committee on Methods of Chemical Analysis, June 12, 1946 [11].

	Maximum Limits of Permissive Variation	
Component	Between two results	Between the extreme values in three results
Silicon dioxide, SiO_2	0.16	0.24
Aluminum oxide, Al_2O_3	0.20	0.30
Ferric oxide, Fe_2O_3	0.10	0.15
Calcium oxide, CaO	0.20	0.30
Magnesium oxide, MgO	0.16	0.24
Sulfur trioxide, SO_3	0.10	0.15
Loss on ignition	0.10	0.15
Sodium oxide, Na_2O	0.03	0.05
Potassium oxide, K_2O	0.03	0.05
Phosphorous pentoxide, P_2O_5	0.03	0.05
Managanic oxide, Mn_2O_3	0.03	0.05
Insoluble residue	0.10	0.15
Chloroform - soluble organic substances	0.004	0.006
Free calcium oxide	0.20	0.30
water – soluble alkali	0.05	0.08

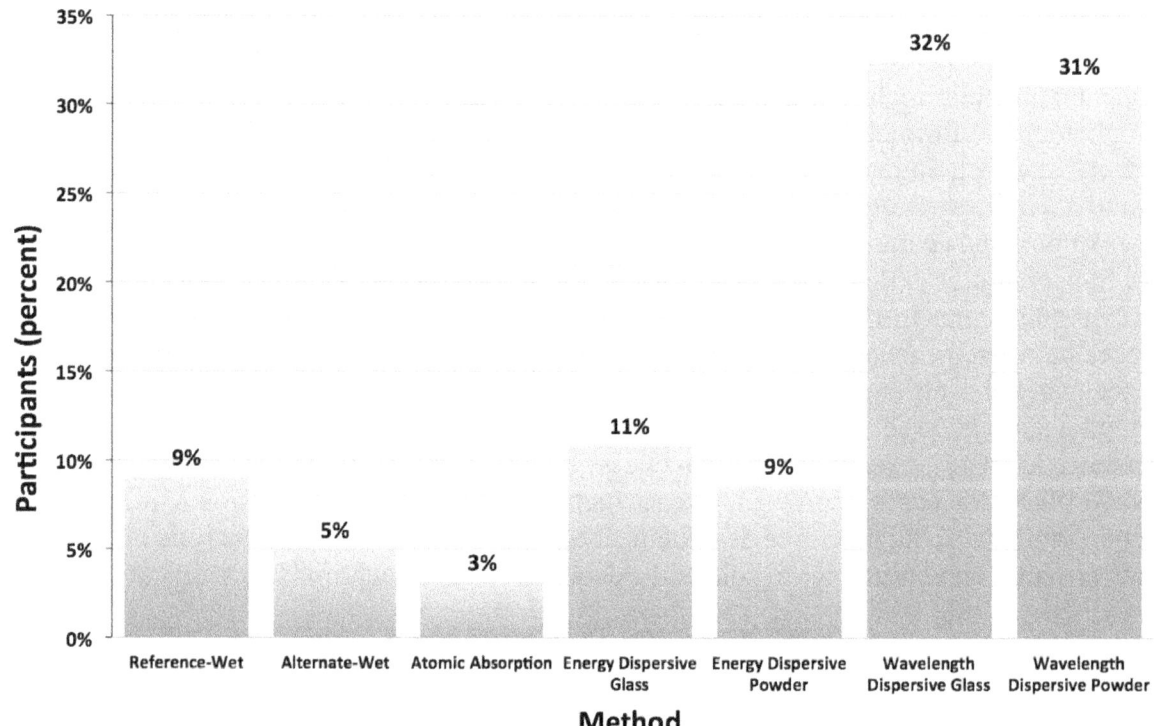

Figure 1 Reference and instrumental methods popularity from CCRL proficiency test data shows that over 80 % of the labs use X-ray methods for bulk chemical analysis by either energy- or wavelength-dispersive analysis, using either powder or glass specimens.

Over 80 % of the participants in the CCRL proficiency test program conduct XRF analysis using either a wavelength- or an energy-dispersive spectrometer [12]. The percentage of cement manufacturers that use XRF analysis is probably higher, and the remaining instrumental and traditional chemical methods of analysis are generally reserved as backup in case the principal XRF instrument is down for repair. In ASTM C114 terminology, the XRF method, atomic absorption and inductively coupled plasma spectrometry fall under rapid methods, whereas much longer times are required to perform the traditional wet, gravimetric, titrimetric, and colorimetric chemical techniques. Current practice for chemical analysis follows ASTM C114, which states that any method of analysis may be used as long as it can be demonstrated to conform to precision and bias performance criteria of Table 1 in ASTM C114 [2]. This means that the rapid methods require qualification where results of six of seven certified reference materials must fall within the qualification criteria of Table 1; the seventh measurement must fall with twice the precision criteria in Table 1. The qualification limits were originally published in 1946 and remain the same today. Bias criteria were originally based upon the maximum difference of three determinations, whereas today the criteria are a maximum limit on the difference of the mean of two determinations and an accepted reference value. The National Bureau of Standards (currently NIST) issued three portland cement reference materials (1011, 1015, and 1016) in January of 1962 with provisional values, which were finalized in 1964. Until this time, no comparison to "certified" values was possible.

ASTM C01.23 initiated an inter-laboratory study for XRF Analysis to establish a data set for assessing precision of the method being developed. This method does not provide a set of instructions to prepare specimens and perform an analysis, but rather outlines a goal of analysis of major and minor elements by XRF with use of either of two specimen preparation procedures: pressed powder and fused glass. Guidelines as a draft standard are provided for both the pressed pellet and the fused glass preparations.

For the inter-laboratory study, three pairs of cements were distributed to participants with approximately 45 laboratories participating. This report covers the second set of cements, referred to as Cements E and F, originating from the CCRL proficiency test program cements 165 and 166, respectively. Mill sheet data report ca. 3 % limestone, with CCRL 165 limestone being 80 % $CaCO_3$ and 166 being 97.4 % $CaCO_3$. Each laboratory was asked to follow their own standard operating procedure (SOP), as long as it fell within the draft standard guidelines, and prepare and analyze three specimens (replicates) to be analyzed twice each (duplicates) on two different days for a total of six specimens and twelve analyses (Appendix A). Subsequent studies will contain ASTM C595 1s cements (samples C D), and ASTM C595 1p cements with fly ash (samples G H). Results and data from each sample set will be reported separately.

Sample preparation for XRF can be achieved using either of two distinct methods: a pressed powder and a fused glass disk. Pressed powder specimens are typically ground in a tungsten carbide ring and puck mill with a binding agent to reduce the particle size and provide a packed powder mount that will remain intact for transport and analysis. The advantages of this preparation method include the simplicity and better detection limits, while disadvantages include the "mineralogical effect", which requires a similar composition matrix between a bracketed calibration and unknown specimens for the calibrations to be valid. The potential for bias due to the mineralogical effect will be greater for the second, third, and fourth cement pairs, which will contain slag additions at the 50 % level and limestone additions at a level of less than 5 %, and fly ash, respectively. The fused disk preparation eliminates the potentially adverse effects of discrete mineral phases by dissolving the cement in a flux and fusing the mixture into a homogeneous glass disk. However, the fluxing process is subject to volatization of some analytes if the heating process is not carefully controlled.

The data analysis follows ASTM E691 [13] "Standard Practice for Conducting an Interlaboratory Study to Determine the Precision of a Test Method," and was performed using the Dataplot[1] software. Data were compiled by the CCRL staff into a database and exported to a spreadsheet format for subsequent processing and analysis. The terms used in ASTM E691 and the means of their calculation are presented in Figure 3, with the last six terms being used subsequently in the evaluation and presentation of the results.

The layout of this report consists of some background for chemical analysis of cements, information on the interlaboratory study, methodology behind the precision calculations, a summary table for all reported elements, and individual analyte (as oxides) results in table and graphical form, along with comparisons to previous studies. The calculations provide precision estimates for consideration by the ASTM C01.23 Subcommittee on compositional analysis for use in developing a draft standard test method for XRF analysis of hydraulic cements. The draft method uses a qualification approach, similar to that used in ASTM C114, having criteria for precision as well as accuracy. Unfortunately, the process and data used to develop the ASTM C114 Table 1 criteria are no longer available, so while precision is calculated here, criteria for accuracy will require careful consideration by the subcommittee. If the predominant sources of uncertainty lie in the laboratory protocol – the sample preparation, the calibrations, and the analyses – the differences between laboratory results reflect the combined within-laboratory uncertainties and laboratory-specific bias based upon protocol. As was done with the XRD test method ASTM C1365, bias limits (ASTM C114, Table 1, Column 3) are established using prediction intervals for each analyte.

---

[1] http://www.itl.nist.gov/div898/software/dataplot/homepage.htm

n = number of test results per cell

p = number of laboratories

x = individual test result

$\bar{x}$ = cell average $\qquad \sum_{i=1}^{n} \bar{x} \Big/ n$

$\overline{X}$ = average of cell averages for one material $\qquad \sum_{i=1}^{p} \bar{x} \Big/ p$

s = cell standard deviation $\qquad \sum_{i=1}^{n} (x - \bar{x})^2 \Big/ (n-1)$

d = cell deviation $\qquad (\bar{x} - \overline{X})$

$S_{\bar{x}}$ = standard deviation of cell averages $\qquad \sum_{i=1}^{p} (d)^2 \Big/ (p-1)$

S_r = repeatability standard deviation $\qquad \sqrt{\sum_{i=1}^{p} s^2 / p}$

s_R = reproducibility standard deviation $\qquad \sqrt{s_{\bar{x}}^2 + s_r^2(n-1)/n}$

h = between-laboratory consistency $\qquad d / s_{\bar{x}}$

k = within-laboratory consistency $\qquad s / s_r$

r = 95 % repeatability statistic $\qquad 2.8 * s_r$

R = 95 % reproducibility statistic $\qquad 2.8 * s_R$

Figure 2 Calculated values for the determination of within and between lab precision.

Measurement Precision

Uncertainties in bulk oxide measurements originate from a number of sources: consistency and bias in specimen preparations, standardization, data collection procedures, and analysis protocol. Measurements are estimates of the actual value being measured and ideally have some statement of uncertainty. The uncertainty may be estimated through an interlaboratory study, which provides estimates on precision, or random error (Type A) and bias, or systematic error (Type B) uncertainty. ASTM defines precision as "the closeness of agreement between independent test results obtained under stipulated conditions", (the standard test procedure), which may be expressed as a standard deviation (1σ) [14]. Precision is further differentiated by that achieved within a laboratory by a single instrument (and operator, or procedure), called repeatability, and that between different laboratories, called reproducibility as a single standard deviation or a 95 % limit as defined by E691 and presented below:

Repeatability: Precision under repeatability conditions

Repeatability limit (r): "The value below which the *absolute difference between two individual test results* obtained under repeatability conditions may be expected to occur with a probability of approximately 0.95 (95 %)"
The repeatability limit is 2.8 (1.96*√2) times the repeatability standard deviation

Reproducibility: Precision under reproducibility conditions

Reproducibility limit (R): The value below which the absolute difference between two test results obtained under reproducibility conditions may be expected to occur with a probability of approximately 0.95 (95 %)
The reproducibility limit is 2.8 (1.96*√2) times the reproducibility standard deviation

Measurement Accuracy and Method Bias

ASTM defines accuracy as "the closeness of agreement between a test result and a accepted reference value" [14], which includes both random and systematic error. The qualification criteria in C 114, Table 1, Column 3 addresses accuracy. Bias is defined by ASTM as "the difference between the expectation of the test results and an accepted reference value" [14], and reflects the systematic error. A meaningful estimate of method bias is more difficult to extract from interlaboratory studies if an explicit protocol is not available, even if certified reference materials are available. Systematic error introduced by individual lab protocols dominates that of the method, making a universal bias correction difficult to estimate and apply. In addition, the cements used in this program were not reference materials (due to the number of participants), so a bias calculation is not possible. These materials were specially homogenized and packaged as part of the CCRL chemical proficiency test program.

Participation in the inter-laboratory study (ILS) was open to all interested laboratories and was not restricted based on the laboratory staff's years of experience. In addition, the ILS did not specify an explicit method for preparation and analysis. These two factors contributed to the overall uncertainty in the measurement data. Some means of identification of outlying data was necessary to exclude any outlying results and their influences on the calculated precision and consensus values. A graphical representation of this process is provided in Figure 4 where the individual replicate data from powder data with results plotted with cement E on the x-axis and cement F on the y-axis. The vertical and horizontal axes represent the consensus value means for cements E and F, respectively for the original data (before outlier identification). Like a Youden plot, this plot of the data pairs provides some insight, based on the degree and directions of dispersion of the results. In the absence of bias, the point pattern is roughly circular, and the dispersion along a diagonal from the lower-left, to the upper-right quadrant represents systematic error, and dispersion that is orthogonal to this direction represents the random error (precision).

In the example for CaO shows powder preparation (squares), glass preparation (circles) and reference methods (+). An X striking out a symbol represents data considered to be an outlier based upon that lab's repeatability or reproducibility. For example, a cluster of circles along the upper-right diagonal exhibits good within-lab precision but significant positive systematic error for both cements, likely representing either a calibration or preparation error. Data scattered perpendicular to the diagonal exhibits systematic error and random error. Finally, on occasion, a lab may report an errant value, which might represent an error in the sample preparation or a data entry error. In this case, no attempt was made to fix the data, and that analyte from that lab was removed for the final analysis to maintain the balanced data set required by E691.

The plots in Figure 5 of the laboratory mean and standard deviation are also useful for rapid visualization of overall performance by laboratory and material. The mean plots represent the mean of the three replicates for cements E and F (labeled as Materials 1 and 2) against the consensus value. The mean values by laboratory have the effect of averaging out the random error in the analyses, providing a more robust estimate for each lab. Lab 1 stands out in producing consistently low values, while precision problems of Lab 3 (cement F, material 2) and Lab 20 (cement E, material 1) are seen in the standard deviation (SD) plots.

Quantitative assessment of within- and between-lab precision is represented by the h and k statistics [13], which can take the table form or be expressed in a plot. The h and k consistency statistics are measures of the lab's within- and between-laboratory precision, are shown in Figure 6. These statistics were used to identify outlying lab data in the original data set, which were subsequently excluded on an analyte-by-analyte basis. The rationale for this being applied on a one time only basis was that it aided in the identification of unusual data due to standardization and procedural error, or errors on the reported values due to data entry. No attempt was made to evaluate the data and fix errant values due to entry error. ASTM E691 requires a balanced data set so if a lab submitted an incomplete data set or was flagged as an outlier due to an errant value, the entire data set for that analyte was eliminated.

Qualification criteria specified in Table 1 of ASTM C114 include limits on the maximum difference between duplicates and an accuracy criterion limiting the maximum difference between the mean of two duplicates and a certificate value from a certified reference material. These criteria are provided here by the within- and between-laboratory precision statistics and by the prediction interval. A prediction interval for a single future observation is an interval having a pre-determined probability that it will "contain the next randomly selected observation from a population" [15]. The prediction interval approach assumes generally that **(1)** the underlying population is normally distributed, is homogeneous and unchanging and, **(2)** that the interval will be used to bracket numbers based on data from that same underlying population. The requirement on the accuracy of the estimates is based upon prediction intervals derived from the composite interlaboratory study data and 2-point means of duplicate samples.

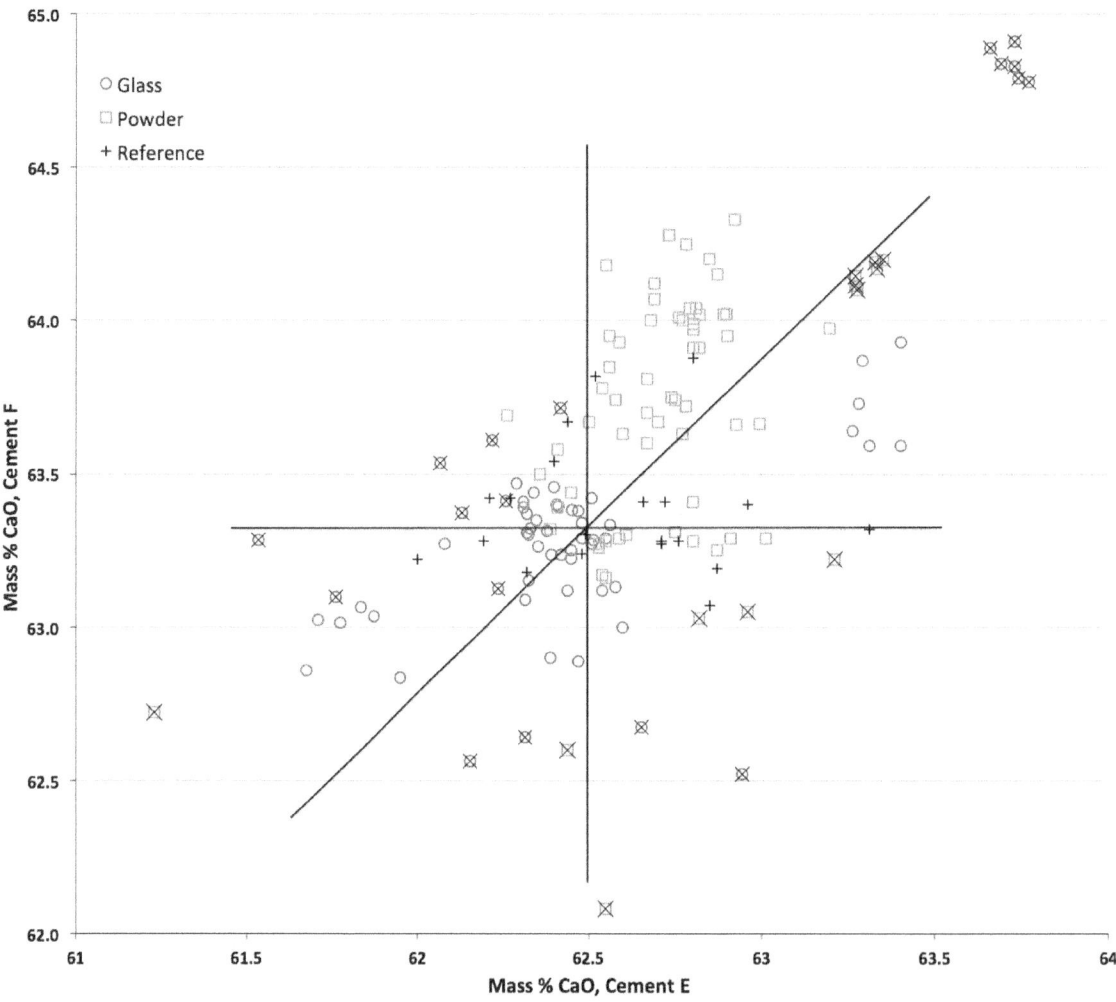

Figure 3 Scatter plot for CaO with XRF-glass as blue circles, XRF-powder as red squares, Reference methods as a +, and x striking out excluded data due to exceeding the *h* or *k* statistics.

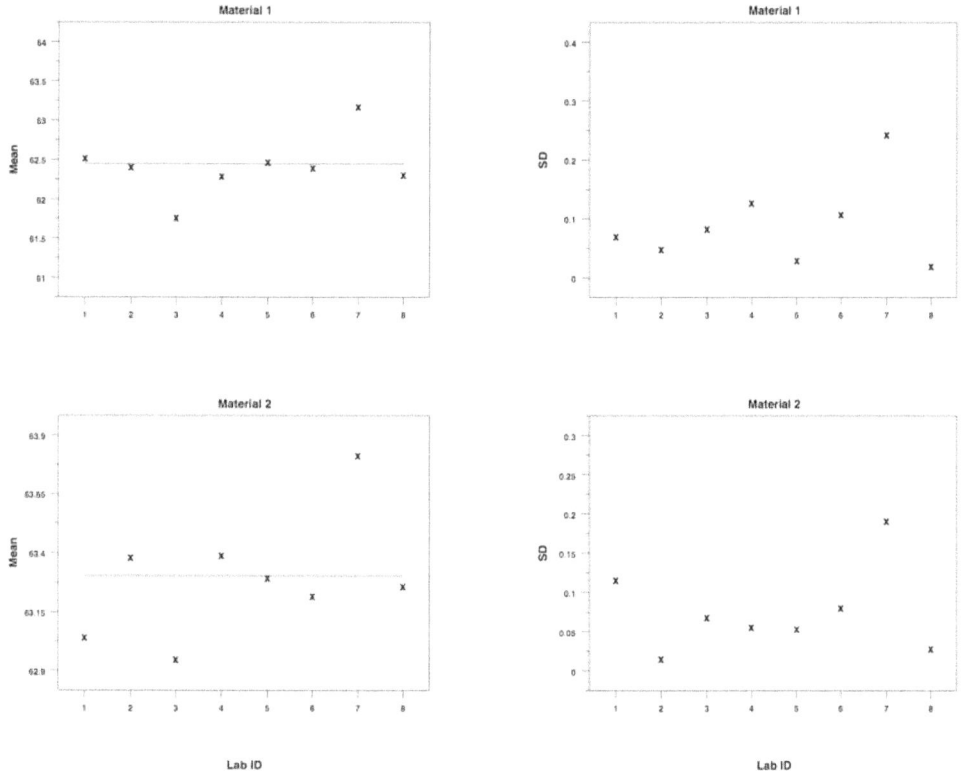

Figure 4 Lab means and standard deviations for cements E (Material 1) and F (material 2).

Results – Condensed Summary

Table 2 summarizes the results by method. The table reports the 1σ within-laboratory standard deviation? (Sr), the 1σ between-laboratory standard deviation (sR), and the appropriate ASTM d2s (the 95 % limits on the difference between two test results) as r and R, representing a pooled standard deviation for the two replicates for both cements E and F for within-laboratory and between-laboratory results, respectively. In addition, pooled results for glass and powder methods are shown on the right-hand side of the table. More detailed summaries by method and material are found in Appendix B for the glass preparation, and in Appendix C for the powder preparation. Appendix D contains the data used in this analysis.

A two-sided 95 % prediction interval for the mean of two test results is presented in Table 2 for each analyte. This interval defines the range of values one might expect relative to a certified value of a reference material based upon the mean of two separate determinations, similar to the criteria of column 3 of Table 1 in ASTM C114, titled "Maximum difference of the average of duplicates from CRM certificate values".

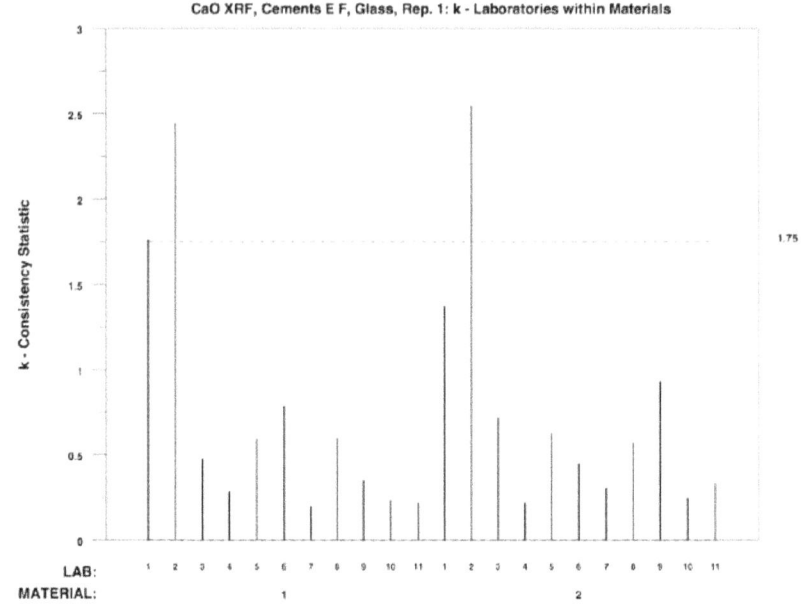

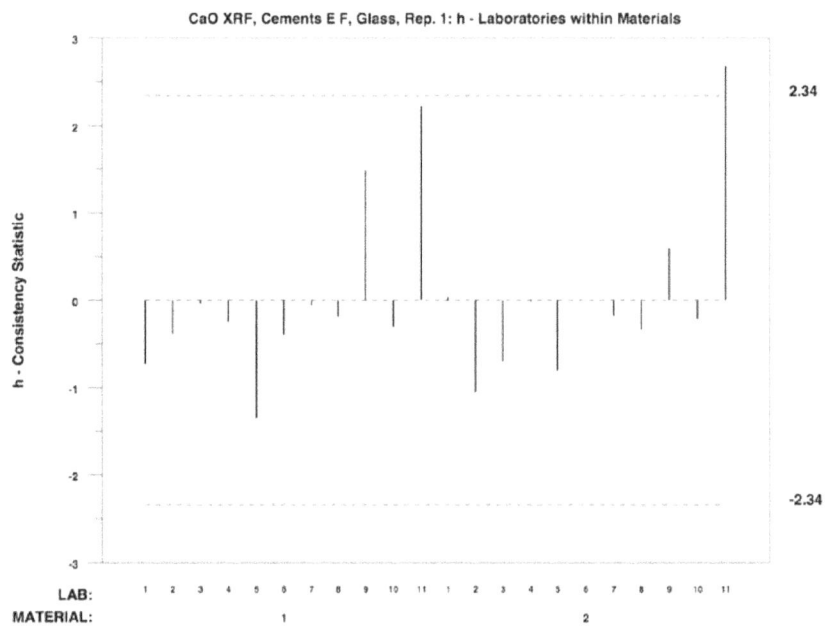

Figure 5 *h* and *k* statistic plots illustrate the within- (*k*) and between-laboratory (*h*) precision and are useful in identifying unusual results that may be considered outlying. The dashed lines mark the limits for each statistic.

Table 2 Pooled Results for XRF-Glass, XRF-Powder and Combined Powder and Glass Methods with ZnO and Cr_2O_3 results from combined glass and powder methods.

	Glass				Powder				Pooled Glass and Powder				
	Sr	sR	r	R	Sr	sR	r	R	Sr		sR	r	R
CaO	0.097	0.346	0.268	0.959	0.136	0.284	0.377	0.787		0.118	0.316	0.327	0.877
SiO$_2$	0.056	0.169	0.154	0.469	0.058	0.251	0.161	0.695		0.057	0.214	0.158	0.593
Al$_2$O$_3$	0.027	0.075	0.074	0.209	0.024	0.099	0.067	0.273		0.025	0.088	0.070	0.243
Fe$_2$O$_3$	0.011	0.039	0.029	0.108	0.013	0.044	0.037	0.122		0.012	0.041	0.033	0.115
SO$_3$	0.047	0.081	0.130	0.225	0.029	0.110	0.079	0.304		0.039	0.097	0.107	0.268
MgO	0.011	0.019	0.031	0.054	0.021	0.046	0.058	0.127		0.017	0.035	0.046	0.098
Na$_2$O	0.006	0.016	0.017	0.044	0.007	0.022	0.019	0.060		0.006	0.019	0.018	0.052
K$_2$O	0.004	0.013	0.011	0.036	0.007	0.018	0.020	0.049		0.006	0.015	0.016	0.043
TiO$_2$	0.004	0.008	0.010	0.023	0.004	0.006	0.010	0.017		0.004	0.007	0.010	0.021
P$_2$O$_5$	0.002	0.004	0.005	0.011	0.003	0.013	0.009	0.035		0.003	0.009	0.007	0.026
Mn$_2$O$_3$	0.002	0.007	0.006	0.019	0.003	0.007	0.007	0.019		0.002	0.006	0.005	0.017
SrO	0.001	0.004	0.002	0.011	0.002	0.006	0.005	0.017		0.001	0.005	0.004	0.014
ZnO										0.0005	0.0042	0.001	0.012
Cr$_2$O$_3$										0.0009	0.0019	0.002	0.005
Cl										0.0016	0.0062	0.005	0.017

Table 3 95 % Prediction Interval designed to bracket values of a mean of k = 2 measurements. The mean of two replicate determinations should differ from the known value of the certified reference material (±) by no more than the value shown.

	Glass	Powder	Pooled
CaO	0.647	0.512	0.501
SiO_2	0.305	0.509	0.375
Al_2O_3	0.132	0.173	0.136
Fe_2O_3	0.068	0.077	0.064
SO_3	0.146	0.188	0.150
MgO	0.036	0.077	0.054
Na_2O	0.028	0.038	0.029
K_2O	0.022	0.031	0.024
TiO_2	0.014	0.010	0.011
P_2O_5	0.007	0.022	0.014
Mn_2O_3	0.014	0.014	0.010
SrO	0.008	0.013	0.008
ZnO			0.007
Cr_2O_3			0.004
Cl			0.013

Each analyte, expressed as oxides, is represented with box plots for both cements E and F that include the reference data from the CCRL proficiency test program for these cements. For comparison, a table of results for each duplicate and replicate by sample preparation (glass and powder), a summary table for the precision calculations by cement, duplicate and replicate, and a bar chart illustrating the pooled results against the ASTM C114 criteria and reproducibility values calculated in other studies on analytical uncertainty.

Box plots are a graphical one-way ANOVA, enabling comparison of the two XRF preparation and the reference methods results through assessment of the alignment or mis-alignment of median values, differences in interquartile ranges, and the extent of the data extremes. The box plots characterize the XRF data after outliers from the initial analysis have been removed.

Important features of the box plot are:

1. the width of each box is proportional to sample size,
2. the median value is identified by the X within the box is used for its resistance to outliers,
3. the interquartile range ("middle half") of the data are represented by the body of the box,
4. the top and the bottom of the box represent the estimated 75 % and 25 % point, respectively, and
5. the extremes (minimum and maximum) are represented by the ends of the straight lines projecting from the box

Summary

Precision and accuracy estimates from analysis of data from an ASTM interlaboratory test program provide the basis for qualification statement for the new XRF standard test method now in development. The current qualification criteria are based upon traditional wet chemistry test methods, and the original data and means of their calculation of the qualification criteria data are not available. Following ASTM E691, precision values for within- and between-laboratory and their 95 % limits have been determined. The accuracy criterion in ASTM C114, Table 1 is developed here using a two-point mean and 95 % prediction interval. Together, these performance criteria will aid in facilitating accurate and consistent analyses of the bulk chemical compositions of hydraulic cements.

Acknowledgements

LeRoy Jacobs, Chair of the XRF task group within C01.23 subcommittee on compositional analysis coordinated the effort to develop a standard test procedure for X-ray fluorescence of hydraulic cements. His leadership, perseverance and humor kept the effort going. Robin Haupt of the Cement and Concrete reference laboratory coordinated the distribution of materials and collection. The comments and suggestions of the internal reviewers, Kenneth Snyder and Clarissa Ferraris and external reviewer, Donald Broton are appreciated.

SiO₂

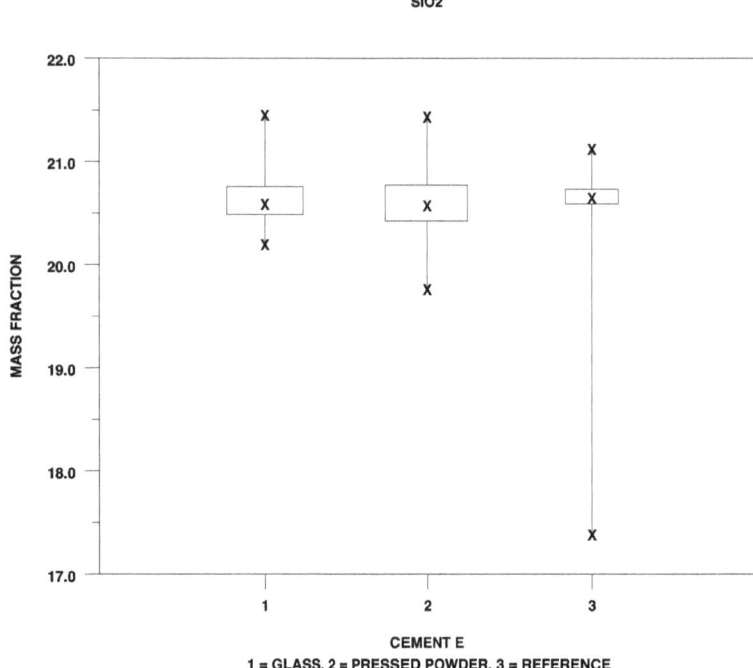

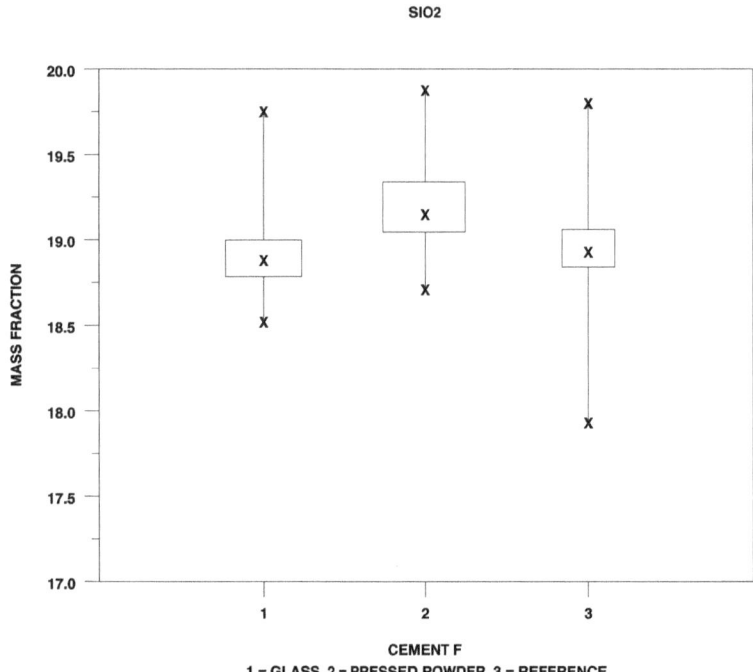

Figure 6 Box plots for SiO₂ for XRF glass and powder, and reference methods.

14

SiO₂, Cement E, Glass, Replicate 1

Laboratory Number	Cell Mean	Cell SD	d	h	k
1	20.4507	0.1215	-0.1219	-0.73	1.97
2	20.8033	0.0450	0.2307	1.37	0.73
3	20.2757	0.0474	-0.2969	-1.77	0.77
4	20.4997	0.0270	-0.0729	-0.43	0.44
5	20.6917	0.0343	0.1191	0.71	0.56
6	20.6883	0.0605	0.1157	0.69	0.98
7	20.5300	0.0329	-0.0426	-0.25	0.53
8	20.7353	0.0904	0.1627	0.97	1.47
9	20.4788	0.0152	-0.0938	-0.56	0.25

```
Average of cell averages              =         20.57261
Standard Deviation of cell averages   =          0.16817
Repeatability Standard Deviation      =          0.06155
Reproducibility Standard Deviation    =          0.17731
Critical Values h,k  =        2.23,  1.73
```

SiO₂, Cement F, Glass, Replicate 1

Laboratory Number	Cell Mean	Cell SD	d	h	k
1	18.7622	0.0983	-0.0789	-0.54	1.93
2	18.9417	0.0479	0.1006	0.69	0.94
3	18.5590	0.0260	-0.2820	-1.94	0.51
4	18.8252	0.0319	-0.0159	-0.11	0.63
5	19.0233	0.0388	0.1823	1.25	0.76
6	18.9217	0.0387	0.0806	0.55	0.76
7	18.7833	0.0197	-0.0577	-0.40	0.39
8	10.0948	0.0677	0.1538	1.06	1.33
9	18.7582	0.0404	-0.0829	-0.57	0.79

```
Average of cell averages              =         18.84104
Standard Deviation of cell averages   =          0.14566
Repeatability Standard Deviation      =          0.05083
Reproducibility Standard Deviation    =          0.15287
Critical Values h,k  =        2.23,  1.73
```

SiO₂, Cement E, Glass, Replicate 2

Laboratory Number	Cell Mean	Cell SD	*d*	*h*	*k*
1	20.4933	0.1167	-0.0883	-0.49	1.99
2	20.8717	0.0671	0.2900	1.62	1.14
3	20.2676	0.0355	-0.3141	-1.75	0.60
4	20.5050	0.0283	-0.0766	-0.43	0.48
5	20.6933	0.0314	0.1117	0.62	0.54
6	20.6683	0.0426	0.0867	0.48	0.73
7	20.5167	0.0234	-0.0650	-0.36	0.40
8	20.7432	0.0845	0.1616	0.90	1.44
9	20.4755	0.0184	-0.1061	-0.59	0.31

```
Average of cell averages          =        20.58162
Standard Deviation of cell averages =       0.17959
Repeatability Standard Deviation  =         0.05868
Reproducibility Standard Deviation =        0.18741
Critical Values h,k  =         2.23, 1.73
```

SiO₂, Cement F, Glass, Replicate 2

Laboratory Number	Cell Mean	Cell SD	*d*	*h*	*k*
1	18.8043	0.0931	-0.0527	-0.35	1.86
2	18.9750	0.0404	0.1180	0.79	0.80
3	18.5647	0.0425	-0.2924	-1.96	0.85
4	18.8480	0.0416	-0.0090	-0.06	0.83
5	19.0217	0.0232	0.1646	1.10	0.46
6	18.9683	0.0500	0.1113	0.75	1.00
7	18.7900	0.0329	-0.0670	-0.45	0.65
8	18.9962	0.0598	0.1391	0.93	1.19
9	18.7450	0.0337	-0.1120	-0.75	0.67

```
Average of cell averages          =        18.85702
Standard Deviation of cell averages =       0.14923
Repeatability Standard Deviation  =         0.05018
Reproducibility Standard Deviation =        0.15610
Critical Values h,k  =         2.23, 1.73
```

SiO$_2$, Cement E, Powder, Replicate 1

Laboratory Number	Cell Mean	Cell SD	d	h	k
1	20.6467	0.0612	0.0891	0.46	0.99
2	20.8367	0.0505	0.2791	1.45	0.81
3	20.7317	0.0232	0.1741	0.90	0.37
4	20.4100	0.0522	-0.1476	-0.77	0.84
5	20.6217	0.0725	0.0641	0.33	1.17
6	20.8417	0.1030	0.2841	1.47	1.66
7	20.5267	0.0437	-0.0309	-0.16	0.70
8	20.3350	0.0572	-0.2226	-1.15	0.92
9	20.4833	0.0388	-0.0742	-0.39	0.63
10	20.4083	0.0436	-0.1492	-0.77	0.70
11	20.2917	0.0914	-0.2659	-1.38	1.47

```
Average of cell averages            =        20.55758
Standard Deviation of cell averages =         0.19280
Repeatability Standard Deviation    =         0.06206
Reproducibility Standard Deviation  =         0.20095
Critical Values h,k  =         2.34, 1.75
```

SiO$_2$, Cement F, Powder, Replicate 1

Laboratory Number	Cell Mean	Cell SD	d	h	k
1	19.0483	0.0183	-0.1063	-0.38	0.37
2	19.1583	0.0786	0.0037	0.01	1.58
3	19.0900	0.0566	-0.0646	-0.23	1.14
4	18.8683	0.0426	-0.2863	-1.02	0.86
5	19.1417	0.0574	-0.0130	-0.05	1.15
6	19.3017	0.0643	0.1470	0.52	1.29
7	19.0533	0.0398	-0.1013	-0.36	0.80
8	19.1333	0.0472	-0.0213	-0.08	0.95
9	18.7467	0.0294	-0.4080	-1.45	0.59
10	19.3183	0.0431	0.1637	0.58	0.87
11	19.8408	0.0410	0.6862	2.44	0.83

```
Average of cell averages            =        19.15462
Standard Deviation of cell averages =         0.28179
Repeatability Standard Deviation    =         0.04971
Reproducibility Standard Deviation  =         0.28542
Critical Values h,k  =         2.34, 1.75
```

SiO$_2$, Cement E, Powder, Replicate 2

Laboratory Number	Cell Mean	Cell SD	*d*	*h*	*k*
1	19.0617	0.0426	−0.1036	−0.35	0.74
2	19.1433	0.0942	−0.0220	−0.07	1.63
3	19.1017	0.0462	−0.0636	−0.21	0.80
4	18.8400	0.0529	−0.3253	−1.09	0.92
5	19.1883	0.0818	0.0230	0.08	1.42
6	19.3167	0.0622	0.1514	0.51	1.08
7	19.0483	0.0214	−0.1170	−0.39	0.37
8	19.1400	0.0400	−0.0253	−0.09	0.69
9	18.7717	0.0279	−0.3936	−1.32	0.48
10	19.2967	0.0301	0.1314	0.44	0.52
11	19.9100	0.0823	0.7447	2.50	1.43

```
Average of cell averages         =         19.16530
Standard Deviation of cell averages  =      0.29760
Repeatability Standard Deviation  =         0.05776
Reproducibility Standard Deviation  =       0.30223

Critical Values h,k  =          2.34, 1.75
```

SiO$_2$, Cement F, Powder, Replicate 2

Laboratory Number	Cell Mean	Cell SD	*d*	*h*	*k*
1	19.0617	0.0426	−0.1036	−0.35	0.74
2	19.1433	0.0942	−0.0220	−0.07	1.63
3	19.1017	0.0462	−0.0636	−0.21	0.80
4	18.8400	0.0529	−0.3253	−1.09	0.92
5	19.1883	0.0818	0.0230	0.08	1.42
6	19.3167	0.0622	0.1514	0.51	1.08
7	19.0483	0.0214	−0.1170	−0.39	0.37
8	19.1400	0.0400	−0.0253	−0.09	0.69
9	18.7717	0.0279	−0.3936	−1.32	0.48
10	19.2967	0.0301	0.1314	0.44	0.52
11	19.9100	0.0823	0.7447	2.50	1.43

```
Average of cell averages         =         19.16530
Standard Deviation of cell averages  =      0.29760
Repeatability Standard Deviation  =         0.05776
Reproducibility Standard Deviation  =       0.30223
Critical Values h,k  =          2.34, 1.75
```

18

XRF Glass, Replicate 1

Material	Xbar	s_x	s_r	s_R	r	R
1	20.5726	0.1682	0.0616	0.1773	0.17	0.50
2	18.8410	0.1457	0.0508	0.1529	0.14	0.43

XRF Glass, Replicate 2

Material	Xbar	s_x	s_r	s_R	r	R
1	20.5816	0.1796	0.0587	0.1874	0.16	0.52
2	18.8570	0.1492	0.0502	0.1561	0.14	0.44

XRF Powder Replicate 1

Material	Xbar	s_x	s_r	s_R	r	R
1	20.5576	0.1928	0.0621	0.2010	0.17	0.56
2	19.1546	0.2818	0.0497	0.2854	0.14	0.80

XRF Powder Replicate 2

Material	Xbar	s_x	s_r	s_R	r	R
1	20.5750	0.1878	0.0620	0.1961	0.17	0.55
2	19.1653	0.2976	0.0578	0.3022	0.16	0.85

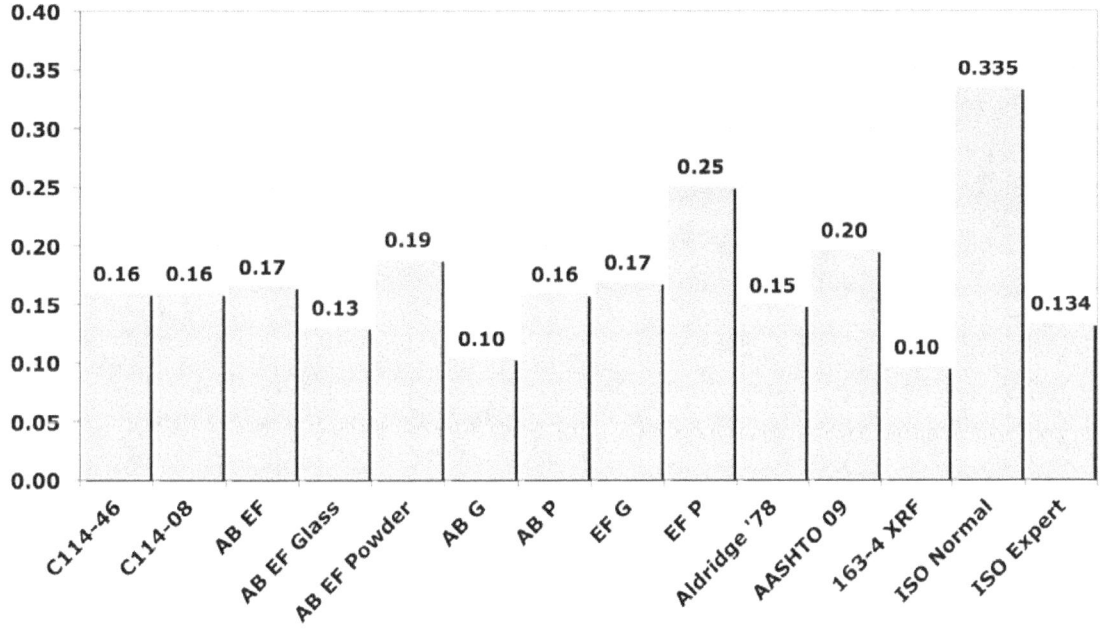

Figure 7 SiO_2 precision statistics by method with bar chart comparing results to current and past ASTM C114 limits and previous studies on chemical analysis precision as 1σ, between lab (S_R).

Al$_2$O$_3$

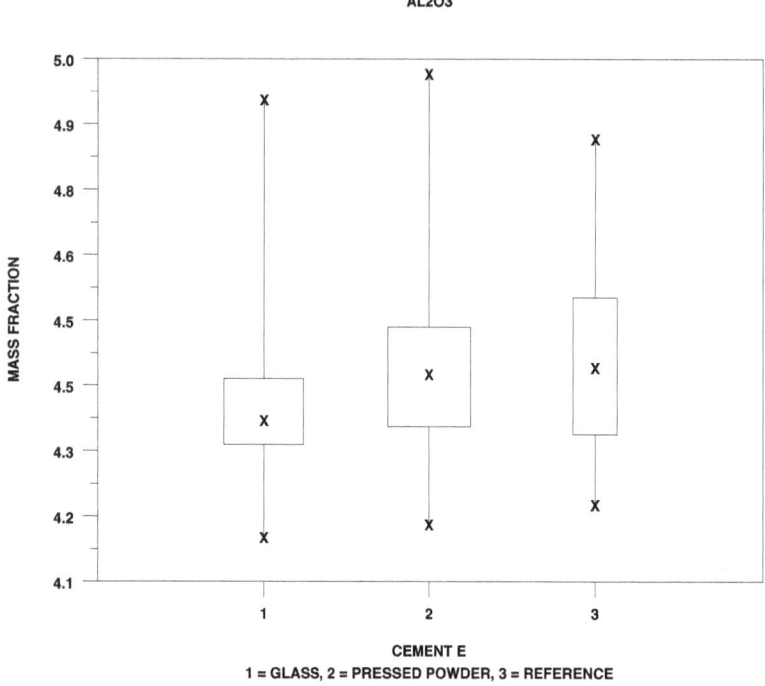

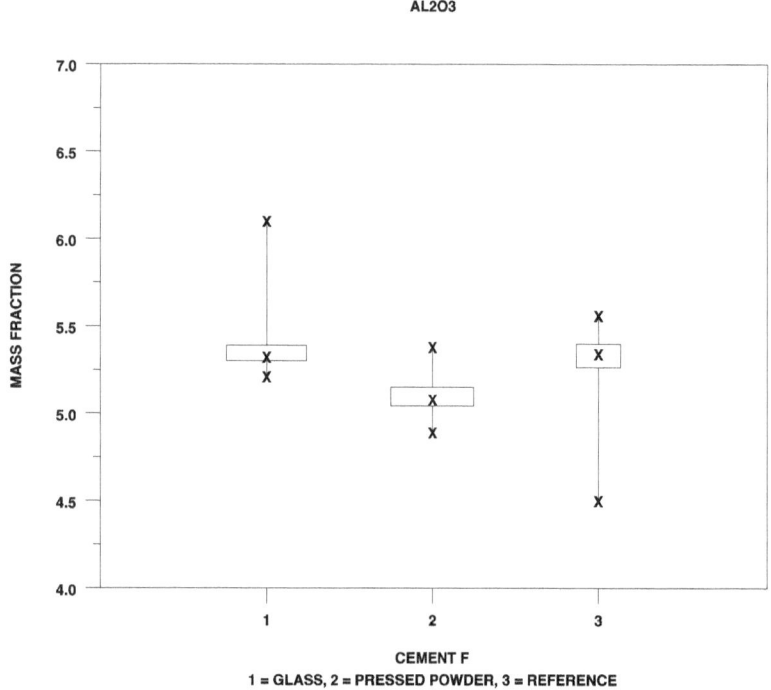

Figure 8 Box plots for Al$_2$O$_3$ for XRF glass and powder, and reference methods.

Al$_2$O$_3$, Cement E, Glass, Rep. 1

Laboratory Number	Cell Mean	Cell SD	d	h	k
1	4.3992	0.0265	-0.0459	-0.56	0.93
2	4.6093	0.0534	0.1643	1.99	1.87
3	4.4967	0.0273	0.0516	0.62	0.96
4	4.4307	0.0107	-0.0144	-0.17	0.37
5	4.4812	0.0267	0.0361	0.44	0.93
6	4.4450	0.0164	0.0000	0.00	0.58
7	4.3017	0.0299	-0.1434	-1.74	1.05
8	4.4869	0.0205	0.0419	0.51	0.72
9	4.3700	0.0352	-0.0750	-0.91	1.23
10	4.4298	0.0143	-0.0152	-0.18	0.50

```
Average of cell averages           =          4.44504
Standard Deviation of cell averages =         0.08263
Repeatability Standard Deviation   =          0.02857
Reproducibility Standard Deviation =          0.08665
Critical Values h,k  =         2.29, 1.74
```

Al$_2$O$_3$, Cement F, Glass, Rep. 1

Laboratory Number	Cell Mean	Cell SD	d	h	k
1	5.2623	0.0263	-0.0741	-1.19	0.96
2	5.4653	0.0474	0.1289	2.07	1.72
3	5.3600	0.0245	0.0236	0.38	0.89
4	5.3027	0.0141	-0.0337	-0.54	0.51
5	5.3692	0.0251	0.0327	0.52	0.91
6	5.3250	0.0138	-0.0115	-0.18	0.50
7	5.2583	0.0240	-0.0781	-1.25	0.87
8	5.3913	0.0231	0.0548	0.88	0.84
9	5.3133	0.0418	-0.0231	-0.37	1.52
10	5.3170	0.0129	-0.0195	-0.31	0.47

```
Average of cell averages           =          5.33645
Standard Deviation of cell averages =         0.06238
Repeatability Standard Deviation   =          0.02753
Reproducibility Standard Deviation =          0.06725
Critical Values h,k  =         2.29, 1.74
```

Al$_2$O$_3$, Cement E, Glass, Rep. 2

Laboratory Number	Cell Mean	Cell SD	d	h	k
1	4.4060	0.0250	-0.0383	-0.50	1.02
2	4.5877	0.0374	0.1433	1.87	1.53
3	4.4933	0.0151	0.0490	0.64	0.62
4	4.4296	0.0151	-0.0148	-0.19	0.62
5	4.4840	0.0141	0.0397	0.52	0.58
6	4.4417	0.0204	-0.0027	-0.03	0.84
7	4.2967	0.0388	-0.1477	-1.93	1.59
8	4.4866	0.0252	0.0423	0.55	1.03
9	4.3950	0.0243	-0.0493	-0.64	0.99
10	4.4228	0.0122	-0.0215	-0.28	0.50

```
Average of cell averages              =       4.44433
Standard Deviation of cell averages   =       0.07652
Repeatability Standard Deviation      =       0.02444
Reproducibility Standard Deviation    =       0.07971
Critical Values h,k  =         2.29, 1.74
```

Al$_2$O$_3$, Cement F, Glass, Rep. 2

Laboratory Number	Cell Mean	Cell SD	d	h	k
1	5.2705	0.0223	-0.0580	-0.94	0.88
2	5.4483	0.0387	0.1198	1.95	1.52
3	5.3533	0.0350	0.0248	0.40	1.38
4	5.2980	0.0099	-0.0305	-0.50	0.39
5	5.3417	0.0153	0.0132	0.21	0.60
6	5.3317	0.0172	0.0032	0.05	0.68
7	5.2333	0.0294	-0.0952	-1.55	1.16
8	5.3929	0.0188	0.0644	1.05	0.74
9	5.2933	0.0356	-0.0352	-0.57	1.40
10	5.3220	0.0111	-0.0065	-0.11	0.44

```
Average of cell averages              =       5.32851
Standard Deviation of cell averages   =       0.06140
Repeatability Standard Deviation      =       0.02542
Reproducibility Standard Deviation    =       0.06564
Critical Values h,k  =         2.29, 1.74
```

Al$_2$O$_3$, Cement E, Powder, Rep. 1

Laboratory Number	Cell Mean	Cell SD	*d*	*h*	*k*
1	4.4700	0.0126	−0.0551	−0.54	0.54
2	4.5433	0.0121	0.0182	0.18	0.51
3	4.5167	0.0103	−0.0084	−0.08	0.44
4	4.5300	0.0477	0.0049	0.05	2.03
5	4.5067	0.0197	−0.0184	−0.18	0.84
6	4.5450	0.0176	0.0199	0.20	0.75
7	4.3900	0.0167	−0.1351	−1.33	0.71
8	4.6117	0.0293	0.0866	0.85	1.24
9	4.3961	0.0144	−0.1290	−1.27	0.61
10	4.7417	0.0284	0.2166	2.12	1.21

```
Average of cell averages           =        4.52511
Standard Deviation of cell averages =        0.10192
Repeatability Standard Deviation   =        0.02354
Reproducibility Standard Deviation =        0.10416
Critical Values h,k  =      2.29, 1.74
```

Al$_2$O$_3$, Cement F, Powder, Rep. 1

Laboratory Number	Cell Mean	Cell SD	*d*	*h*	*k*
1	5.1300	0.0167	0.0115	0.12	0.66
2	5.3517	0.0240	0.2332	2.41	0.95
3	5.1733	0.0308	0.0549	0.57	1.22
4	5.0783	0.0366	−0.0401	−0.41	1.45
5	5.0400	0.0179	−0.0785	−0.81	0.71
6	5.0333	0.0207	−0.0851	−0.88	0.82
7	5.0467	0.0121	−0.0718	−0.74	0.48
8	5.0733	0.0163	−0.0451	−0.47	0.65
9	5.1762	0.0376	0.0577	0.60	1.49
10	5.0818	0.0253	−0.0367	−0.38	1.00

```
Average of cell averages           =        5.11847
Standard Deviation of cell averages =        0.09685
Repeatability Standard Deviation   =        0.02522
Reproducibility Standard Deviation =        0.09954
Critical Values h,k  =      2.29, 1.74
```

Al$_2$O$_3$, Cement E, Powder, Rep. 2

Laboratory Number	Cell Mean	Cell SD	d	h	k
1	4.4483	0.0286	-0.0670	-0.76	1.22
2	4.5433	0.0151	0.0280	0.32	0.64
3	4.5117	0.0098	-0.0037	-0.04	0.42
4	4.5317	0.0527	0.0163	0.18	2.25
5	4.5033	0.0163	-0.0120	-0.14	0.70
6	4.5367	0.0186	0.0213	0.24	0.80
7	4.3883	0.0172	-0.1270	-1.44	0.74
8	4.6200	0.0110	0.1046	1.19	0.47
9	4.4011	0.0137	-0.1142	-1.30	0.59
10	4.6692	0.0183	0.1538	1.74	0.78

```
Average of cell averages            =        4.51537
Standard Deviation of cell averages =        0.08815
Repeatability Standard Deviation    =        0.02338
Reproducibility Standard Deviation  =        0.09070
Critical Values h,k  =        2.29, 1.74
```

Al$_2$O$_3$, Cement F, Powder, Rep. 2

Laboratory Number	Cell Mean	Cell SD	d	h	k
1	5.1433	0.0225	0.0158	0.16	0.91
2	5.3567	0.0258	0.2291	2.36	1.05
3	5.1783	0.0133	0.0508	0.52	0.54
4	5.0833	0.0408	-0.0442	-0.46	1.65
5	5.0400	0.0126	-0.0875	-0.90	0.51
6	5.0317	0.0160	-0.0959	-0.99	0.65
7	5.0483	0.0117	-0.0792	-0.82	0.47
8	5.0833	0.0207	-0.0442	-0.46	0.84
9	5.1770	0.0363	0.0495	0.51	1.47
10	5.1333	0.0279	0.0058	0.06	1.13

```
Average of cell averages            =        5.12753
Standard Deviation of cell averages =        0.09709
Repeatability Standard Deviation    =        0.02468
Reproducibility Standard Deviation  =        0.09967
Critical Values h,k  =        2.29, 1.74
```

XRF Glass, Replicate 1

Material	Xbar	s_x	s_r	s_R	r	R
1	4.4450	0.0826	0.0286	0.0866	0.08	0.24
2	5.3365	0.0624	0.0275	0.0672	0.08	0.19

XRF Glass, Replicate 2

Material	Xbar	s_x	s_r	s_R	r	R
1	4.4443	0.0765	0.0244	0.0797	0.07	0.22
2	5.3285	0.0614	0.0254	0.0656	0.07	0.18

XRF Powder Replicate 1

Material	Xbar	s_x	s_r	s_R	r	R
1	4.5154	0.0882	0.0234	0.0907	0.07	0.25
2	5.1275	0.0971	0.0247	0.0997	0.07	0.28

XRF Powder Replicate 2

Material	Xbar	s_x	s_r	s_R	r	R
1	4.5251	0.1019	0.0235	0.1042	0.07	0.29
2	5.1185	0.0968	0.0252	0.0995	0.07	0.28

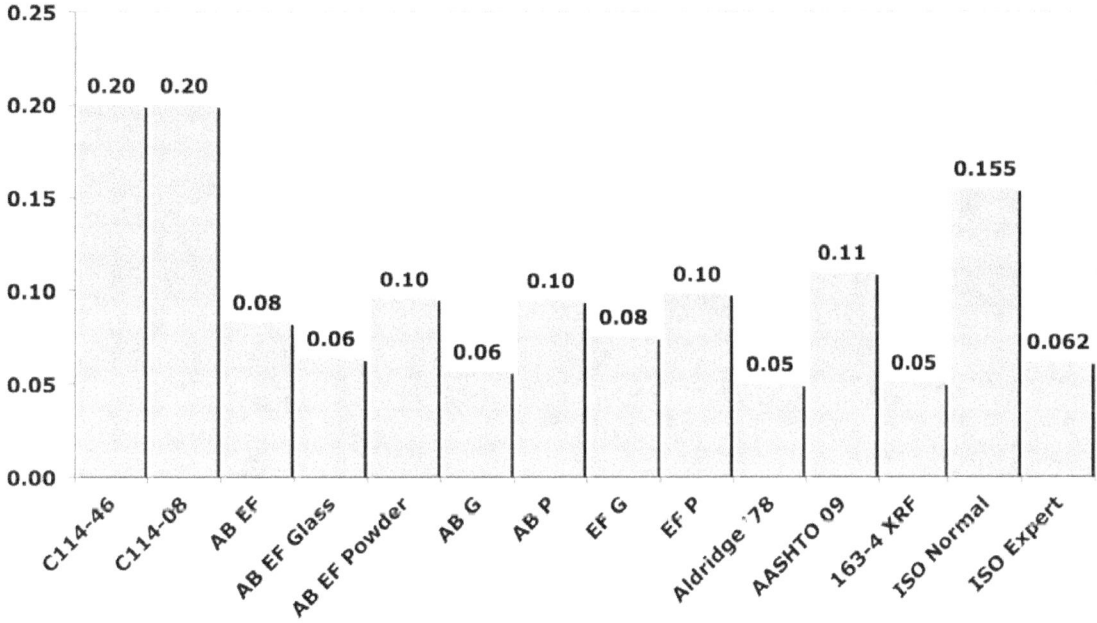

Figure 9 Al_2O_3 precision statistics by method with bar chart comparing results to current and past ASTM C114 limits and previous studies on chemical analysis precision as 1σ, between lab (S_R).

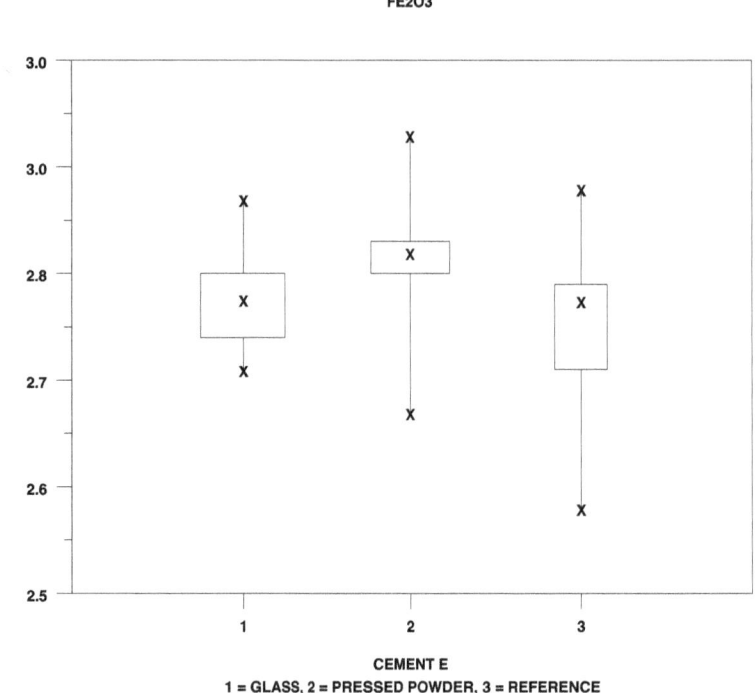

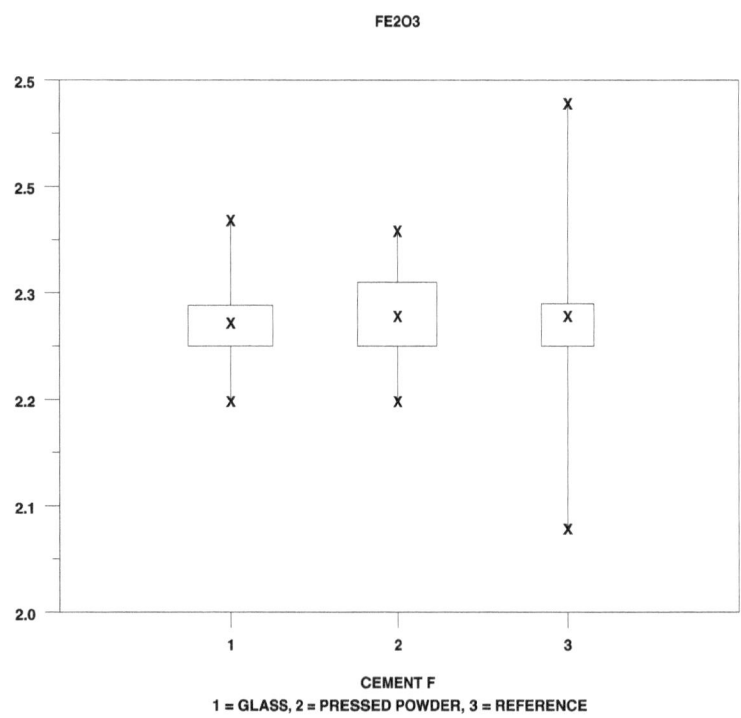

Figure 10 Box plots for Fe$_2$O$_3$ for XRF glass and powder, and reference methods.

Fe$_2$O$_3$, Cements E, Glass, Rep. 1

Laboratory Number	Cell Mean	Cell SD	d	h	k
1	2.8658	0.0071	-0.0191	-0.49	0.96
2	2.9133	0.0140	0.0284	0.73	1.87
3	2.8795	0.0045	-0.0054	-0.14	0.60
4	2.8373	0.0067	-0.0476	-1.22	0.90
5	2.8317	0.0075	-0.0533	-1.37	1.01
6	2.8983	0.0041	0.0134	0.34	0.55
7	2.8883	0.0098	0.0034	0.09	1.32
8	2.8936	0.0067	0.0087	0.22	0.90
9	2.9683	0.0041	0.0834	2.14	0.55
10	2.8730	0.0038	-0.0119	-0.31	0.51

```
Average of cell averages           =        2.88492
Standard Deviation of cell averages =        0.03897
Repeatability Standard Deviation   =        0.00747
Reproducibility Standard Deviation =        0.00747
Critical Values h,k  =        2.29, 1.74
```

Fe$_2$O$_3$, Cements F, Glass, Rep. 1

Laboratory Number	Cell Mean	Cell SD	d	h	k
1	2.3672	0.0114	-0.0605	-0.39	0.03
2	2.3897	0.0115	-0.0380	-0.24	0.03
3	2.3736	0.0027	-0.0541	-0.35	0.01
4	2.3475	0.0062	-0.0801	-0.51	0.02
5	2.3333	0.0137	-0.0943	-0.61	0.04
6	2.3900	0.0063	-0.0376	-0.24	0.02
7	2.8600	1.1856	0.4324	2.78	3.16
8	2.3872	0.0073	-0.0404	-0.26	0.02
9	2.4600	0.0063	0.0324	0.21	0.02
10	2.3678	0.0077	-0.0598	-0.38	0.02

```
Average of cell averages           =        2.42763
Standard Deviation of cell averages =        0.15562
Repeatability Standard Deviation   =        0.37501
Reproducibility Standard Deviation =        0.37501
Critical Values h,k  =        2.29, 1.74
```

Fe$_2$O$_3$, Cements E, Glass, Rep. 2

Laboratory Number	Cell Mean	Cell SD	d	h	k
1	2.8758	0.0080	-0.0095	-0.26	0.86
2	2.9120	0.0138	0.0266	0.72	1.49
3	2.8795	0.0054	-0.0058	-0.16	0.58
4	2.8360	0.0097	-0.0494	-1.34	1.05
5	2.8283	0.0117	-0.0570	-1.55	1.27
6	2.8983	0.0041	0.0130	0.35	0.44
7	2.8933	0.0052	0.0080	0.22	0.56
8	2.8939	0.0066	0.0086	0.23	0.72
9	2.9583	0.0147	0.0730	1.98	1.60
10	2.8780	0.0059	-0.0074	-0.20	0.64

```
Average of cell averages          =       2.88536
Standard Deviation of cell averages  =    0.03683
Repeatability Standard Deviation  =       0.00922
Reproducibility Standard Deviation  =     0.03778
Critical Values h,k  =        2.29, 1.74
```

Fe$_2$O$_3$, Cements F, Glass, Rep. 2

Laboratory Number	Cell Mean	Cell SD	d	h	k
1	2.3762	0.0068	-0.0032	-0.10	0.75
2	2.3898	0.0114	0.0104	0.31	1.24
3	2.3739	0.0046	-0.0055	-0.16	0.50
4	2.3405	0.0105	-0.0389	-1.15	1.15
5	2.3317	0.0160	-0.0478	-1.41	1.75
6	2.3933	0.0052	0.0139	0.41	0.56
7	2.3750	0.0055	-0.0044	-0.13	0.60
8	2.3873	0.0079	0.0079	0.23	0.86
9	2.4567	0.0082	0.0773	2.29	0.89
10	2.3698	0.0094	-0.0096	-0.28	1.03

```
Average of cell averages          =       2.37942
Standard Deviation of cell averages  =    0.03379
Repeatability Standard Deviation  =       0.00916
Reproducibility Standard Deviation  =     0.03481
Critical Values h,k  =        2.29, 1.74
```

Fe$_2$O$_3$, Cements E, Powder, Rep. 1

Laboratory Number	Cell Mean	Cell SD	d	h	k
1	2.9117	0.0098	-0.0098	-0.24	0.70
2	2.8867	0.0082	-0.0348	-0.85	0.58
3	2.9200	0.0126	-0.0015	-0.04	0.91
4	2.9233	0.0216	0.0018	0.04	1.55
5	2.8767	0.0082	-0.0448	-1.10	0.58
6	2.8967	0.0121	-0.0248	-0.61	0.87
7	2.9283	0.0075	0.0068	0.17	0.54
8	2.9367	0.0103	0.0152	0.37	0.74
9	2.9102	0.0276	-0.0114	-0.28	1.98
10	3.0250	0.0045	0.1035	2.53	0.32

```
Average of cell averages          =        2.92151
Standard Deviation of cell averages =      0.04091
Repeatability Standard Deviation   =       0.01396
Reproducibility Standard Deviation =       0.04285
Critical Values h,k  =       2.29, 1.74
```

Fe$_2$O$_3$, Cements F, Powder, Rep. 1

Laboratory Number	Cell Mean	Cell SD	d	h	k
1	2.3867	0.0082	-0.0007	-0.02	0.64
2	2.4000	0.0063	0.0126	0.28	0.50
3	2.4033	0.0151	0.0160	0.36	1.18
4	2.4483	0.0117	0.0610	1.37	0.92
5	2.3033	0.0082	-0.0840	-1.88	0.64
6	2.3517	0.0075	-0.0357	-0.80	0.59
7	2.3717	0.0098	-0.0157	-0.35	0.77
8	2.3767	0.0103	-0.0107	-0.24	0.81
9	2.3746	0.0284	-0.0128	-0.29	2.23
10	2.4575	0.0042	0.0701	1.57	0.33

```
Average of cell averages          =        2.38737
Standard Deviation of cell averages =      0.04461
Repeatability Standard Deviation   =       0.01272
Reproducibility Standard Deviation =       0.04610
Critical Values h,k  =       2.29, 1.74
```

29

Fe$_2$O$_3$, Cements E, Powder, Rep. 2

Laboratory Number	Cell Mean	Cell SD	d	h	k
1	2.9100	0.0089	-0.0108	-0.30	0.67
2	2.8950	0.0084	-0.0258	-0.71	0.63
3	2.9133	0.0082	-0.0074	-0.20	0.62
4	2.9250	0.0187	0.0042	0.12	1.41
5	2.8767	0.0103	-0.0441	-1.22	0.78
6	2.8983	0.0133	-0.0224	-0.62	1.00
7	2.9300	0.0000	0.0092	0.25	0.00
8	2.9400	0.0089	0.0192	0.53	0.67
9	2.9093	0.0284	-0.0115	-0.32	2.14
10	3.0100	0.0045	0.0892	2.46	0.34

```
Average of cell averages            =        2.92076
Standard Deviation of cell averages =        0.03629
Repeatability Standard Deviation    =        0.01325
Reproducibility Standard Deviation  =        0.03825
Critical Values h,k  =       2.29, 1.74
```

Fe$_2$O$_3$, Cements F, Powder, Rep. 2

Laboratory Number	Cell Mean	Cell SD	d	h	k
1	2.3917	0.0117	0.0032	0.07	0.88
2	2.3967	0.0082	0.0082	0.18	0.61
3	2.4050	0.0138	0.0165	0.35	1.03
4	2.4483	0.0117	0.0598	1.29	0.88
5	2.3000	0.0000	-0.0885	-1.90	0.00
6	2.3550	0.0105	-0.0335	-0.72	0.79
7	2.3733	0.0103	-0.0152	-0.33	0.77
8	2.3767	0.0103	-0.0118	-0.25	0.77
9	2.3726	0.0276	-0.0159	-0.34	2.07
10	2.4658	0.0128	0.0773	1.66	0.96

```
Average of cell averages            =        2.38851
Standard Deviation of cell averages =        0.04653
Repeatability Standard Deviation    =        0.01334
Reproducibility Standard Deviation  =        0.04810
Critical Values h,k  =       2.29, 1.74
```

30

XRF Glass, Replicate 1

Material	Xbar	s_x	s_r	s_R	r	R
	2.8849	0.0390	0.0075	0.0396	0.02	0.11
	2.4276	0.1556	0.3750	0.3760	1.05	1.05

XRF Glass, Replicate 2

Material	Xbar	s_x	s_r	s_R	r	R
1	2.8854	0.0368	0.0092	0.0378	0.03	0.11
2	2.3794	0.0338	0.0092	0.0348	0.03	0.10

XRF Powder Replicate 1

Material	Xbar	s_x	s_r	s_R	r	R
1	2.9215	0.0409	0.0140	0.0428	0.04	0.12
2	2.3874	0.0446	0.0127	0.0461	0.04	0.13

XRF Powder Replicate 2

Material	Xbar	s_x	s_r	s_R	r	R
1	2.9208	0.0363	0.0133	0.0383	0.04	0.11
2	2.3885	0.0465	0.0133	0.0481	0.04	0.13

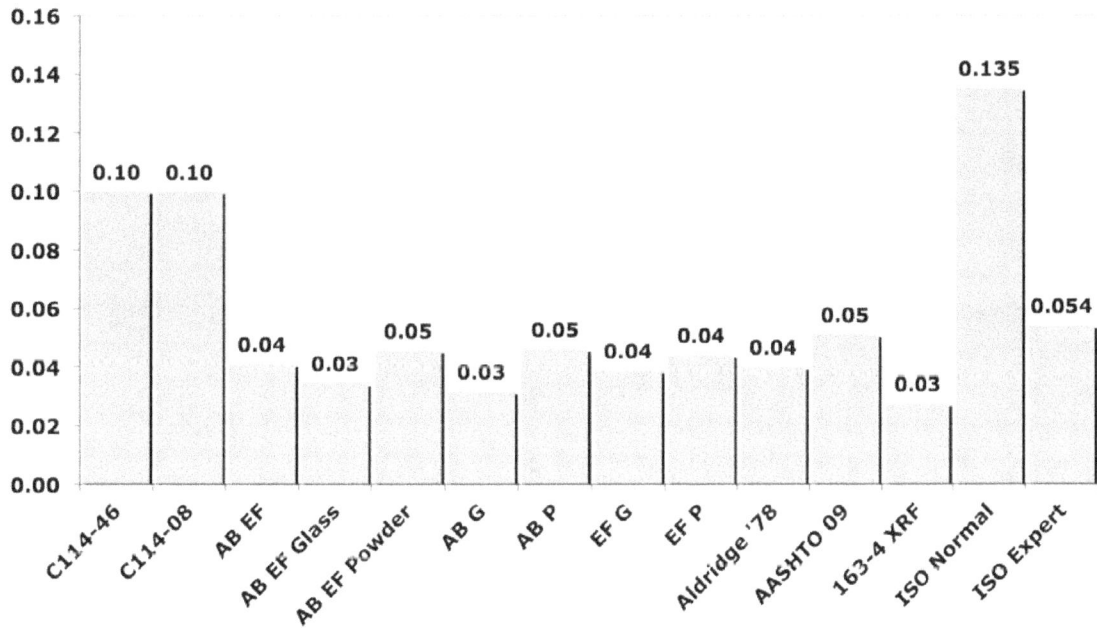

Fe2O3

Figure 11 Fe_2O_3 precision statistics by method with bar chart comparing results to current and past ASTM C114 limits and previous studies on chemical analysis precision as 1σ, between lab (S_R).

CAO

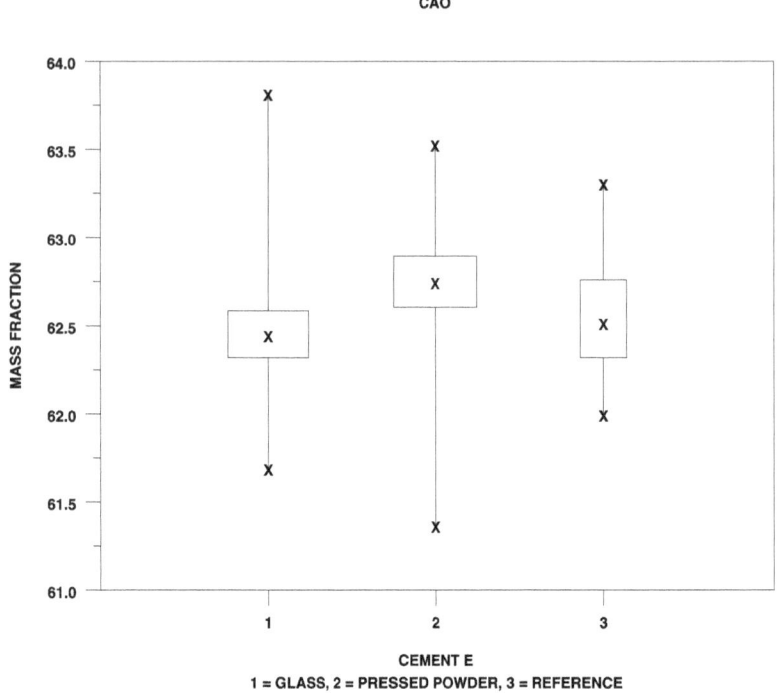

CEMENT E
1 = GLASS, 2 = PRESSED POWDER, 3 = REFERENCE

CAO

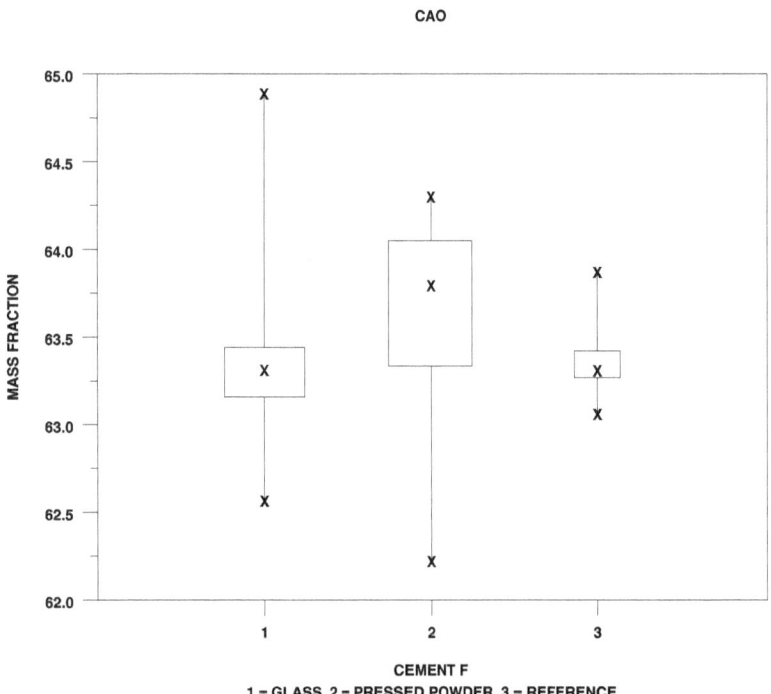

CEMENT F
1 = GLASS, 2 = PRESSED POWDER, 3 = REFERENCE

Figure 12 Box plots for CaO for XRF glass and powder, and reference methods.

CaO, Cements E, Glass, Rep. 1

Laboratory Number	Cell Mean	Cell SD	d	h	k
1	62.5033	0.0831	0.0540	0.13	0.99
2	62.3887	0.0507	-0.0606	-0.14	0.60
3	61.7987	0.1038	-0.6506	-1.56	1.23
4	62.3083	0.1373	-0.1410	-0.34	1.63
5	62.4900	0.0352	0.0407	0.10	0.42
6	62.4203	0.1047	-0.0290	-0.07	1.24
7	63.3233	0.0615	0.8740	2.09	0.73
8	62.3617	0.0413	-0.0876	-0.21	0.49

```
Average of cell averages           =       62.44929
Standard Deviation of cell averages =       0.41814
Repeatability Standard Deviation    =       0.08427
Reproducibility Standard Deviation  =       0.42516
Critical Values h,k  =        2.15, 1.72
```

CaO, Cements F, Glass, Rep. 1

Laboratory Number	Cell Mean	Cell SD	d	h	k
1	63.0267	0.1127	-0.2645	-1.13	1.28
2	63.3952	0.0348	0.1040	0.44	0.40
3	62.9720	0.0983	-0.3191	-1.36	1.12
4	63.3967	0.0703	0.1055	0.45	0.80
5	63.3033	0.0480	0.0122	0.05	0.55
6	63.2237	0.0898	-0.0675	-0.29	1.02
7	63.7250	0.1461	0.4339	1.85	1.66
8	63.2865	0.0390	-0.0046	-0.02	0.44

```
Average of cell averages           =       63.29113
Standard Deviation of cell averages =       0.23504
Repeatability Standard Deviation    =       0.08788
Reproducibility Standard Deviation  =       0.24835
Critical Values h,k  =        2.15, 1.72
```

CaO, Cements E, Glass, Rep. 2

Laboratory Number	Cell Mean	Cell SD	d	h	k
1	62.5467	0.0734	0.1011	0.26	0.63
2	62.4373	0.0524	-0.0082	-0.02	0.45
3	61.7898	0.0866	-0.6557	-1.70	0.75
4	62.3200	0.1311	-0.1255	-0.33	1.13
5	62.5050	0.0333	0.0595	0.15	0.29
6	62.4277	0.1119	-0.0179	-0.05	0.96
7	63.2017	0.2470	0.7561	1.96	2.13
8	62.3362	0.0235	-0.1094	-0.28	0.20

```
Average of cell averages           =        62.44554
Standard Deviation of cell averages =         0.38611
Repeatability Standard Deviation   =         0.11621
Reproducibility Standard Deviation =         0.40042
Critical Values h,k  =         2.15, 1.72
```

CaO, Cements F, Glass, Rep. 2

Laboratory Number	Cell Mean	Cell SD	d	h	k
1	63.0500	0.1182	-0.2526	-0.97	1.25
2	63.3920	0.0184	0.0894	0.34	0.19
3	62.9587	0.0713	-0.3440	-1.32	0.75
4	63.3967	0.0589	0.0940	0.36	0.62
5	63.3050	0.0561	0.0024	0.01	0.59
6	63.2273	0.0837	-0.0753	-0.29	0.88
7	63.8233	0.1941	0.5207	1.99	2.05
8	63.2682	0.0314	-0.0345	-0.13	0.33

```
Average of cell averages           =        63.30265
Standard Deviation of cell averages =         0.26103
Repeatability Standard Deviation   =         0.09464
Reproducibility Standard Deviation =         0.27496
Critical Values h,k  =         2.15, 1.72
```

CaO, Cements E, Powder, Rep. 1

Laboratory Number	Cell Mean	Cell SD	d	h	k
1	62.7733	0.0513	0.0832	0.52	0.48
2	62.6883	0.0886	-0.0018	-0.01	0.84
3	62.7000	0.1166	0.0099	0.06	1.10
4	62.8567	0.0942	0.1665	1.04	0.89
5	62.4550	0.1126	-0.2351	-1.47	1.06
6	62.5067	0.1407	-0.1835	-1.14	1.33
7	62.5617	0.0313	-0.1285	-0.80	0.30
8	62.7267	0.1218	0.0365	0.23	1.15
9	62.9428	0.1397	0.2527	1.58	1.32

```
Average of cell averages            =        62.69013
Standard Deviation of cell averages =         0.16023
Repeatability Standard Deviation    =         0.10579
Reproducibility Standard Deviation  =         0.18709
Critical Values h,k  =        2.23, 1.73
```

CaO, Cements F, Powder, Rep. 1

Laboratory Number	Cell Mean	Cell SD	d	h	k
1	63.9883	0.0445	0.2820	0.89	0.32
2	63.7433	0.1109	0.0370	0.12	0.80
3	63.8533	0.1768	0.1470	0.46	1.28
4	63.3050	0.0550	-0.4013	-1.26	0.40
5	63.5650	0.1790	-0.1413	-0.44	1.30
6	63.5483	0.1783	-0.1580	-0.50	1.29
7	63.2433	0.0622	-0.4630	-1.45	0.45
8	64.2050	0.0993	0.4987	1.57	0.72
9	63.9053	0.2128	0.1990	0.62	1.54

```
Average of cell averages            =        63.70633
Standard Deviation of cell averages =         0.31842
Repeatability Standard Deviation    =         0.13801
Reproducibility Standard Deviation  =         0.34244
Critical Values h,k  =        2.23, 1.73
```

CaO, Cements E, Powder, Rep. 2

Laboratory Number	Cell Mean	Cell SD	d	h	k
1	62.7783	0.0854	0.0656	0.42	0.53
2	62.6617	0.0659	-0.0511	-0.33	0.41
3	62.7150	0.1219	0.0022	0.01	0.75
4	62.8017	0.0674	0.0889	0.57	0.42
5	62.4300	0.0940	-0.2828	-1.80	0.58
6	62.5317	0.1093	-0.1811	-1.15	0.68
7	62.7317	0.3928	0.0189	0.12	2.43
8	62.8067	0.1015	0.0939	0.60	0.63
9	62.9582	0.1378	0.2454	1.56	0.85

```
Average of cell averages              =        62.71276
Standard Deviation of cell averages   =         0.15691
Repeatability Standard Deviation      =         0.16173
Reproducibility Standard Deviation    =         0.21545
Critical Values h,k  =         2.23, 1.73
```

CaO, Cements F, Powder, Rep. 2

Laboratory Number	Cell Mean	Cell SD	d	h	k
1	64.0567	0.0631	0.3331	1.01	0.47
2	63.7850	0.0729	0.0614	0.19	0.55
3	63.8700	0.1774	0.1464	0.44	1.33
4	63.3083	0.0337	-0.4153	-1.26	0.25
5	63.5283	0.1735	-0.1953	-0.59	1.30
6	63.5983	0.1972	-0.1253	-0.38	1.48
7	63.2400	0.0613	-0.4836	-1.46	0.46
8	64.2167	0.0763	0.4931	1.49	0.57
9	63.9090	0.1967	0.1854	0.56	1.48

```
Average of cell averages              =        63.72359
Standard Deviation of cell averages   =         0.33049
Repeatability Standard Deviation      =         0.13298
Reproducibility Standard Deviation    =         0.35208
Critical Values h,k  =         2.23, 1.73
```

XRF Glass, Replicate 1

Material	Xbar	s_x	s_r	s_R	r	R
1	62.4493	0.4181	0.0843	0.4252	0.24	1.19
2	63.2911	0.2350	0.0879	0.2484	0.25	0.70

XRF Glass, Replicate 2

Material	Xbar	s_x	s_r	s_R	r	R
1	62.4455	0.3861	0.1162	0.4004	0.33	1.12
2	63.3026	0.2610	0.0946	0.2750	0.27	0.77

XRF Powder Replicate 1

Material	Xbar	s_x	s_r	s_R	r	R
1	62.6901	0.1602	0.1058	0.1871	0.30	0.52
2	63.7063	0.3184	0.1380	0.3424	0.39	0.96

XRF Powder Replicate 1

Material	Xbar	s_x	s_r	s_R	r	R
1	62.7128	0.1569	0.1617	0.2154	0.45	0.60
2	63.7236	0.3305	0.1330	0.3521	0.37	0.99

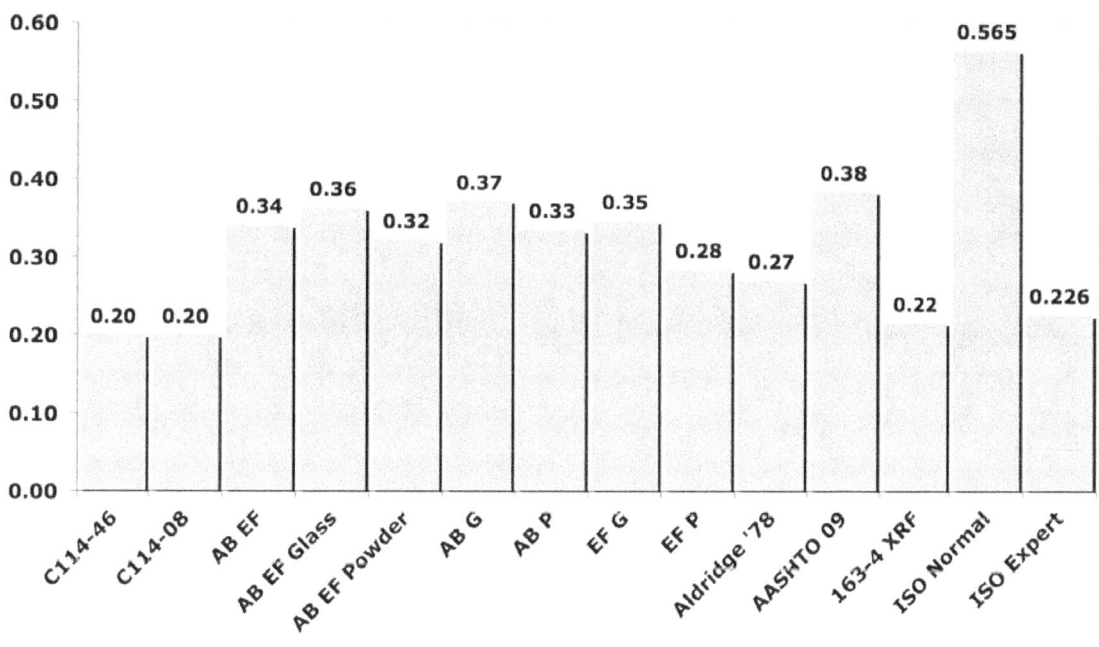

Figure 13 CaO precision statistics by method with bar chart comparing results to current and past ASTM C114 limits and previous studies on chemical analysis precision as 1σ, between lab (S_R).

MgO

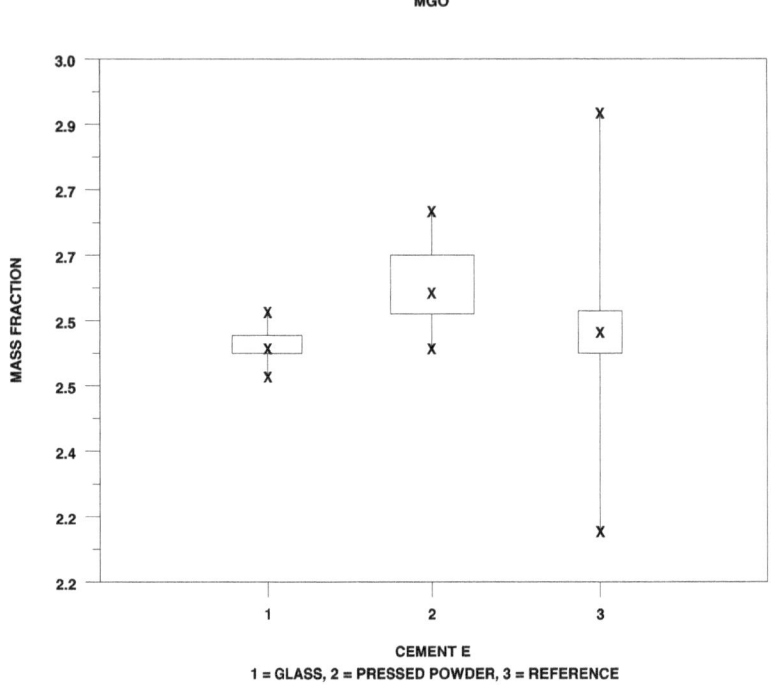

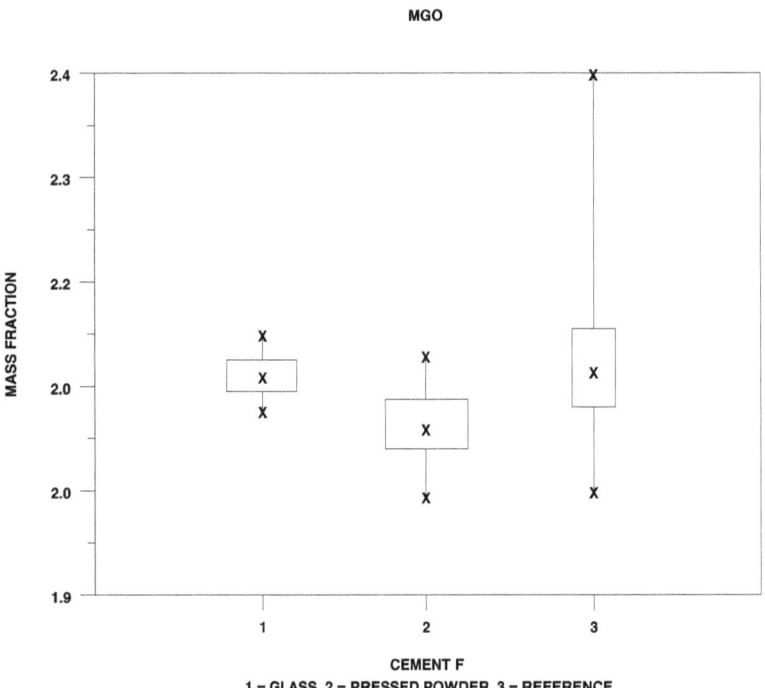

Figure 14 Box plots for MgO for XRF glass and powder, and reference methods.

MgO, Cements E, Glass, Rep. 1

Laboratory Number	Cell Mean	Cell SD	d	h	k
1	2.5422	0.0138	-0.0223	-1.30	1.05
2	2.5683	0.0133	0.0039	0.22	1.01
3	2.5685	0.0056	0.0040	0.23	0.42
4	2.5483	0.0075	-0.0161	-0.94	0.57
5	2.5783	0.0133	0.0139	0.81	1.01
6	2.5583	0.0075	-0.0061	-0.36	0.57
7	2.5744	0.0259	0.0099	0.58	1.97
8	2.5468	0.0033	-0.0176	-1.03	0.25
9	2.5950	0.0138	0.0305	1.78	1.05

```
Average of cell averages           =        2.56447
Standard Deviation of cell averages =       0.01718
Repeatability Standard Deviation   =        0.01316
Reproducibility Standard Deviation  =       0.02096
Critical Values h,k  =          2.23, 1.73
```

MgO, Cements F, Glass, Rep. 1

Laboratory Number	Cell Mean	Cell SD	d	h	k
1	2.1110	0.0083	0.0018	0.11	0.87
2	2.1150	0.0105	0.0058	0.36	1.10
3	2.1055	0.0051	-0.0037	-0.23	0.53
4	2.0900	0.0063	-0.0192	-1.19	0.67
5	2.1350	0.0084	0.0258	1.60	0.88
6	2.0933	0.0137	-0.0158	-0.98	1.44
7	2.1253	0.0080	0.0161	1.00	0.84
8	2.0892	0.0085	-0.0200	-1.24	0.89
9	2.1183	0.0133	0.0092	0.57	1.40

```
Average of cell averages           =        2.10917
Standard Deviation of cell averages =       0.01612
Repeatability Standard Deviation   =        0.00950
Reproducibility Standard Deviation  =       0.01831
Critical Values h,k  =          2.23, 1.73
```

MgO, Cements E, Glass, Rep. 2

Laboratory Number	Cell Mean	Cell SD	d	h	k
1	2.5463	0.0128	-0.0184	-1.00	1.12
2	2.5683	0.0117	0.0036	0.20	1.03
3	2.5699	0.0114	0.0052	0.28	1.00
4	2.5483	0.0075	-0.0164	-0.89	0.66
5	2.5800	0.0089	0.0153	0.83	0.79
6	2.5450	0.0105	-0.0197	-1.07	0.92
7	2.5756	0.0184	0.0109	0.59	1.62
8	2.5507	0.0098	-0.0141	-0.76	0.86
9	2.5983	0.0075	0.0336	1.83	0.66

```
Average of cell averages          =         2.56472
Standard Deviation of cell averages  =      0.01842
Repeatability Standard Deviation  =         0.01138
Reproducibility Standard Deviation  =       0.02115
Critical Values h,k  =        2.23, 1.73
```

MgO, Cements F, Glass, Rep. 2

Laboratory Number	Cell Mean	Cell SD	d	h	k
1	2.1168	0.0053	0.0064	0.44	0.53
2	2.1200	0.0141	0.0095	0.67	1.40
3	2.1038	0.0062	-0.0066	-0.46	0.62
4	2.0900	0.0000	-0.0205	-1.43	0.00
5	2.1333	0.0121	0.0229	1.60	1.20
6	2.1000	0.0110	-0.0105	-0.73	1.09
7	2.1144	0.0122	0.0040	0.28	1.21
8	2.0942	0.0074	-0.0163	-1.14	0.73
9	2.1217	0.0133	0.0112	0.78	1.32

```
Average of cell averages          =         2.11048
Standard Deviation of cell averages  =      0.01429
Repeatability Standard Deviation  =         0.01008
Reproducibility Standard Deviation  =       0.01699
Critical Values h,k  =        2.23, 1.73
```

MgO, Cements E, Powder, Rep. 1

Laboratory Number	Cell Mean	Cell SD	d	h	k
1	2.6483	0.0382	-0.0038	-0.07	1.47
2	2.5700	0.0089	-0.0821	-1.56	0.35
3	2.6617	0.0371	0.0095	0.18	1.43
4	2.6417	0.0117	-0.0105	-0.20	0.45
5	2.5917	0.0172	-0.0605	-1.15	0.67
6	2.6000	0.0200	-0.0521	-0.99	0.77
7	2.7067	0.0367	0.0545	1.03	1.42
8	2.6350	0.0187	-0.0171	-0.32	0.72
9	2.7000	0.0190	0.0479	0.91	0.73
10	2.7417	0.0360	0.0895	1.70	1.39
11	2.6232	0.0274	-0.0289	-0.55	1.06
12	2.7058	0.0139	0.0537	1.02	0.54

```
Average of cell averages            =        2.65214
Standard Deviation of cell averages =        0.05277
Repeatability Standard Deviation    =        0.02589
Reproducibility Standard Deviation  =        0.05782
Critical Values h,k  =        2.38, 1.76
```

MgO, Cements F, Powder, Rep. 1

Laboratory Number	Cell Mean	Cell SD	d	h	k
1	2.0467	0.0258	-0.0152	-0.51	1.70
2	2.1067	0.0137	0.0448	1.51	0.90
3	2.0483	0.0133	-0.0135	-0.46	0.87
4	2.0467	0.0151	-0.0152	-0.51	0.99
5	2.0467	0.0082	-0.0152	-0.51	0.54
6	2.0633	0.0137	0.0015	0.05	0.90
7	2.0500	0.0167	-0.0119	-0.40	1.10
8	2.0517	0.0194	-0.0102	-0.34	1.28
9	2.0783	0.0147	0.0165	0.56	0.97
10	2.1050	0.0084	0.0431	1.45	0.55
11	2.0948	0.0163	0.0329	1.11	1.07
12	2.0042	0.0074	-0.0577	-1.95	0.48

```
Average of cell averages            =        2.06186
Standard Deviation of cell averages =        0.02966
Repeatability Standard Deviation    =        0.01521
Reproducibility Standard Deviation  =        0.03274
Critical Values h,k  =        2.38, 1.76
```

MgO, Cements E, Powder, Rep. 2

Laboratory Number	Cell Mean	Cell SD	d	h	k
1	2.6283	0.0214	-0.0212	-0.42	0.90
2	2.5700	0.0089	-0.0795	-1.56	0.38
3	2.6800	0.0316	0.0305	0.60	1.33
4	2.6417	0.0098	-0.0079	-0.15	0.41
5	2.5933	0.0207	-0.0562	-1.10	0.87
6	2.6050	0.0207	-0.0445	-0.87	0.87
7	2.7017	0.0331	0.0521	1.02	1.39
8	2.6417	0.0183	-0.0079	-0.15	0.77
9	2.7000	0.0167	0.0505	0.99	0.70
10	2.7467	0.0403	0.0971	1.91	1.69
11	2.6236	0.0307	-0.0259	-0.51	1.29
12	2.6625	0.0061	0.0130	0.25	0.26

```
Average of cell averages           =        2.64954
Standard Deviation of cell averages =        0.05091
Repeatability Standard Deviation    =        0.02382
Reproducibility Standard Deviation  =        0.05536
Critical Values h,k  =        2.38, 1.76
```

MgO, Cements F, Powder, Rep. 2

Laboratory Number	Cell Mean	Cell SD	d	h	k
1	2.0650	0.0152	-0.0002	-0.01	0.94
2	2.1050	0.0164	0.0398	1.51	1.01
3	2.0533	0.0216	-0.0119	-0.45	1.33
4	2.0467	0.0082	-0.0185	-0.70	0.50
5	2.0450	0.0197	-0.0202	-0.77	1.22
6	2.0633	0.0137	-0.0019	-0.07	0.84
7	2.0483	0.0133	-0.0169	-0.64	0.82
8	2.0567	0.0207	-0.0085	-0.32	1.28
9	2.0750	0.0122	0.0098	0.37	0.76
10	2.1067	0.0137	0.0415	1.58	0.84
11	2.0967	0.0153	0.0315	1.20	0.95
12	2.0208	0.0191	-0.0444	-1.69	1.18

```
Average of cell averages           =        2.06521
Standard Deviation of cell averages =        0.02630
Repeatability Standard Deviation    =        0.01620
Reproducibility Standard Deviation  =        0.03018
Critical Values h,k  =        2.38, 1.76
```

XRF Glass, Replicate 1

Material	Xbar	s_x	s_r	s_R	r	R
1	2.5645	0.0172	0.0132	0.0210	0.04	0.06
2	2.1092	0.0161	0.0095	0.0183	0.03	0.05

XRF Glass, Replicate 2

Material	Xbar	s_x	s_r	s_R	r	R
1	2.5647	0.0184	0.0114	0.0211	0.03	0.06
2	2.1105	0.0143	0.0101	0.0170	0.03	0.05

XRF Powder Replicate 1

Material	Xbar	s_x	s_r	s_R	r	R
1	2.6521	0.0528	0.0259	0.0578	0.07	0.16
2	2.0619	0.0297	0.0152	0.0327	0.04	0.09

XRF Powder Replicate 2

Material	Xbar	s_x	s_r	s_R	r	R
1	2.6495	0.0509	0.0238	0.0554	0.07	0.15
2	2.0652	0.0263	0.0162	0.0302	0.05	0.08

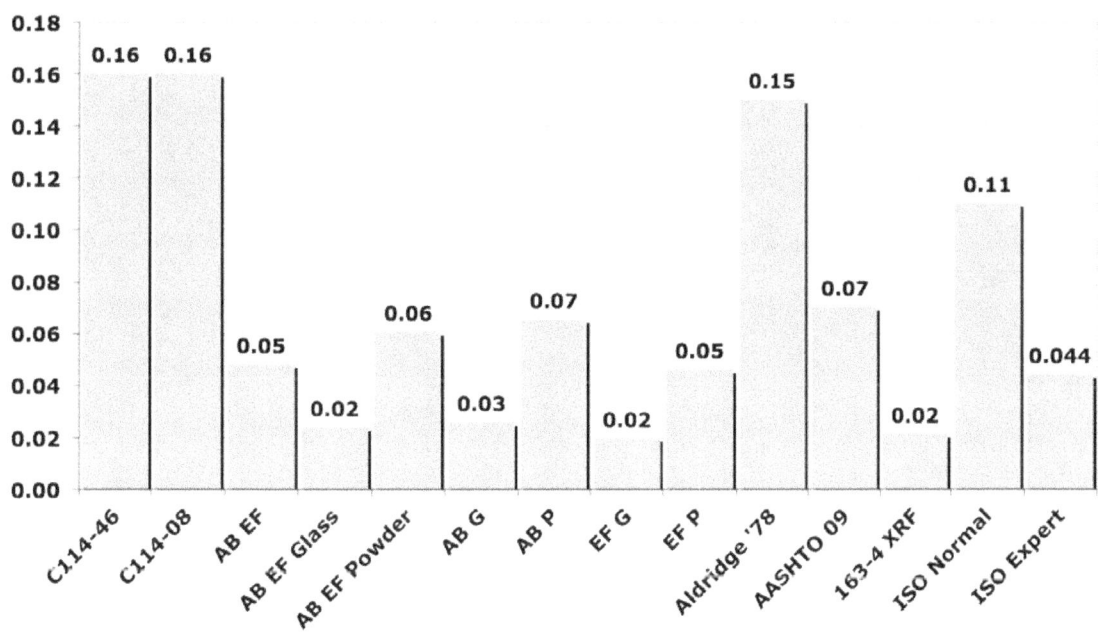

Figure 15 MgO precision statistics by method with bar chart comparing results to current and past ASTM C114 limits and previous studies on chemical analysis precision as 1σ, between lab (S_R).

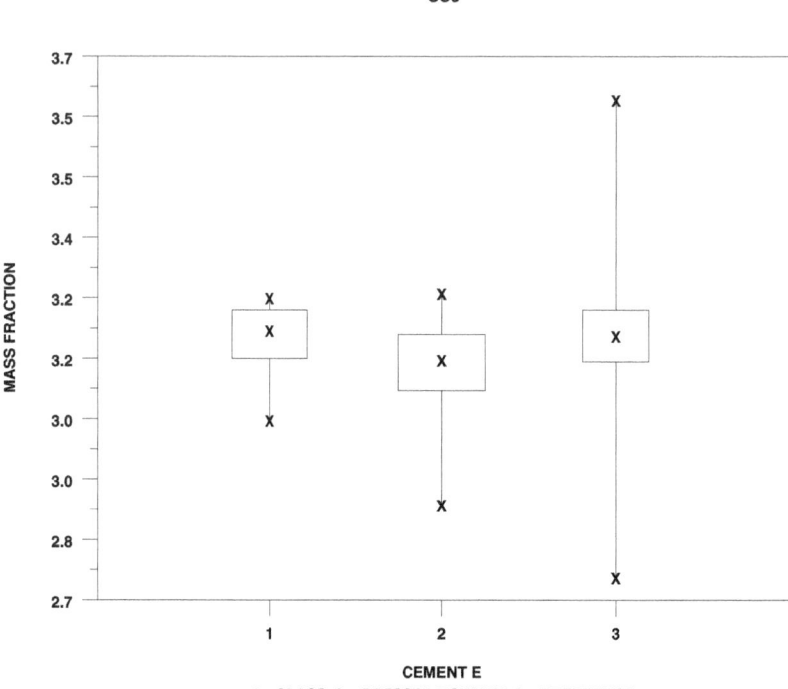

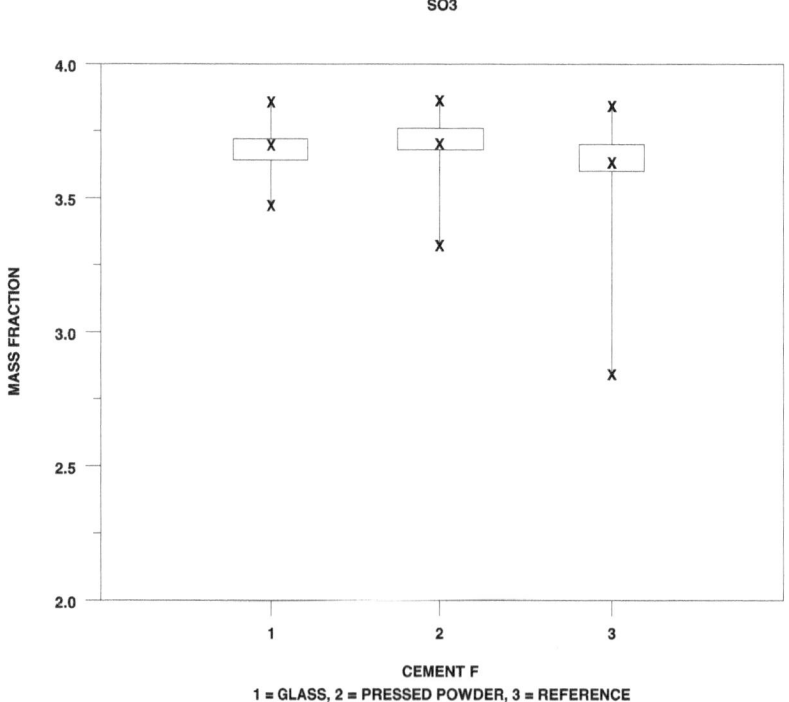

Figure 16 Box plots for SO$_3$ for XRF glass and powder, and reference methods.

SO$_3$, Cements E, Glass, Rep. 1

Laboratory Number	Cell Mean	Cell SD	d	h	k
1	3.2318	0.0211	-0.0036	-0.07	1.63
2	3.2450	0.0182	0.0095	0.18	1.41
3	3.2867	0.0103	0.0512	0.94	0.80
4	3.2942	0.0062	0.0587	1.08	0.48
5	3.2183	0.0133	-0.0171	-0.31	1.03
6	3.2550	0.0055	0.0195	0.36	0.42
7	3.1785	0.0086	-0.0570	-1.05	0.66
8	3.1283	0.0160	-0.1071	-1.97	1.24
9	3.2813	0.0070	0.0459	0.84	0.54

```
Average of cell averages          =        3.23547
Standard Deviation of cell averages  =     0.05440
Repeatability Standard Deviation   =       0.01294
Reproducibility Standard Deviation  =      0.05567
Critical Values h,k  =       2.23,  1.73
```

SO$_3$, Cements F, Glass, Rep. 1

Laboratory Number	Cell Mean	Cell SD	d	h	k
1	3.6952	0.0851	0.0130	0.16	2.64
2	3.6830	0.0306	0.0008	0.01	0.95
3	3.7183	0.0160	0.0362	0.45	0.50
4	3.7220	0.0085	0.0399	0.50	0.26
5	3.6150	0.0122	-0.0672	-0.84	0.38
6	3.7083	0.0117	0.0262	0.33	0.36
7	3.7889	0.0098	0.1067	1.33	0.30
8	3.5050	0.0197	-0.1772	-2.21	0.61
9	3.7038	0.0099	0.0217	0.27	0.31

```
Average of cell averages          =        3.68218
Standard Deviation of cell averages  =     0.08024
Repeatability Standard Deviation   =       0.03228
Reproducibility Standard Deviation  =      0.08548
Critical Values h,k  =       2.23,  1.73
```

SO$_3$, Cements E, Glass, Rep.2

Laboratory Number	Cell Mean	Cell SD	d	h	k
1	3.2390	0.0205	0.0051	0.10	1.49
2	3.2278	0.0177	-0.0060	-0.11	1.29
3	3.2883	0.0147	0.0545	1.02	1.07
4	3.2960	0.0066	0.0621	1.17	0.48
5	3.2150	0.0187	-0.0189	-0.35	1.36
6	3.2533	0.0103	0.0195	0.37	0.75
7	3.1789	0.0094	-0.0550	-1.03	0.68
8	3.1317	0.0098	-0.1022	-1.92	0.71
9	3.2747	0.0086	0.0408	0.77	0.63

```
Average of cell averages          =        3.23386
Standard Deviation of cell averages  =        0.05334
Repeatability Standard Deviation  =        0.01378
Reproducibility Standard Deviation  =        0.05480
Critical Values h,k  =        2.23, 1.73
```

SO$_3$, Cements F, Glass, Rep. 2

Laboratory Number	Cell Mean	Cell SD	d	h	k
1	3.6765	0.0175	0.0074	0.09	0.20
2	3.5815	0.2541	-0.0876	-1.07	2.97
3	3.7183	0.0098	0.0492	0.60	0.11
4	3.7194	0.0072	0.0503	0.61	0.08
5	3.6133	0.0163	-0.0558	-0.68	0.19
6	3.7033	0.0082	0.0342	0.42	0.10
7	3.7898	0.0090	0.1207	1.47	0.10
8	3.5217	0.0232	-0.1474	-1.79	0.27
9	3.6980	0.0104	0.0289	0.35	0.12

```
Average of cell averages          =        3.66910
Standard Deviation of cell averages  =        0.08224
Repeatability Standard Deviation  =        0.08570
Reproducibility Standard Deviation  =        0.11350
Critical Values h,k  =        2.23, 1.73
```

SO$_3$, Cements E, Powder, Rep. 1

Laboratory Number	Cell Mean	Cell SD	d	h	k
1	3.2233	0.0163	0.0412	0.47	0.70
2	2.9800	0.0110	-0.2021	-2.32	0.47
3	3.2117	0.0319	0.0295	0.34	1.37
4	3.1783	0.0147	-0.0038	-0.04	0.63
5	3.2867	0.0258	0.1045	1.20	1.11
6	3.2400	0.0522	0.0579	0.66	2.23
7	3.0967	0.0121	-0.0855	-0.98	0.52
8	3.2217	0.0133	0.0395	0.45	0.57
9	3.2667	0.0103	0.0845	0.97	0.44
10	3.1593	0.0168	-0.0229	-0.26	0.72
11	3.1392	0.0166	-0.0430	-0.49	0.71

```
Average of cell averages              =        3.18213
Standard Deviation of cell averages   =        0.08724
Repeatability Standard Deviation      =        0.02334
Reproducibility Standard Deviation    =        0.08980
Critical Values h,k   =          2.34, 1.75
```

SO$_3$, Cements F, Powder, Rep. 1

Laboratory Number	Cell Mean	Cell SD	d	h	k
1	3.7033	0.0121	0.0028	0.02	0.37
2	3.3900	0.0434	-0.3105	-2.68	1.34
3	3.6867	0.0393	-0.0138	-0.12	1.21
4	3.6800	0.0219	-0.0205	-0.18	0.67
5	3.8067	0.0516	0.1062	0.91	1.59
6	3.8367	0.0175	0.1362	1.17	0.54
7	3.7217	0.0232	0.0212	0.18	0.71
8	3.7517	0.0147	0.0512	0.44	0.45
9	3.7633	0.0288	0.0628	0.54	0.89
10	3.6930	0.0205	-0.0075	-0.07	0.63
11	3.6725	0.0507	-0.0280	-0.24	1.56

```
Average of cell averages              =        3.70050
Standard Deviation of cell averages   =        0.11606
Repeatability Standard Deviation      =        0.03247
Reproducibility Standard Deviation    =        0.11978
Critical Values h,k   =          2.34, 1.75
```

SO$_3$, Cements E, Powder, Rep. 2

Laboratory Number	Cell Mean	Cell SD	d	h	k
1	3.2317	0.0147	0.0615	0.60	0.64
2	2.9783	0.0098	-0.1918	-1.87	0.42
3	3.2017	0.0319	0.0315	0.31	1.38
4	3.1833	0.0052	0.0132	0.13	0.22
5	3.2917	0.0256	0.1215	1.18	1.11
6	3.2333	0.0575	0.0632	0.61	2.49
7	3.1017	0.0117	-0.0685	-0.67	0.51
8	3.2267	0.0103	0.0565	0.55	0.45
9	3.2667	0.0103	0.0965	0.94	0.45
10	3.1526	0.0143	-0.0175	-0.17	0.62
11	3.0042	0.0038	-0.1660	-1.62	0.16

```
Average of cell averages          =        3.17016
Standard Deviation of cell averages  =     0.10274
Repeatability Standard Deviation   =       0.02314
Reproducibility Standard Deviation  =      0.10489
Critical Values h,k  =        2.34, 1.75
```

SO$_3$, Cements F, Powder, Rep. 2

Laboratory Number	Cell Mean	Cell SD	d	h	k
1	3.7133	0.0082	0.0131	0.11	0.24
2	3.3917	0.0431	-0.3085	-2.62	1.28
3	3.6883	0.0417	-0.0119	-0.10	1.24
4	3.6883	0.0256	-0.0119	-0.10	0.76
5	3.8083	0.0500	0.1081	0.92	1.49
6	3.8417	0.0147	0.1415	1.20	0.44
7	3.7183	0.0204	0.0181	0.15	0.61
8	3.7567	0.0121	0.0565	0.48	0.36
9	3.7633	0.0250	0.0631	0.54	0.74
10	3.6924	0.0213	-0.0078	-0.07	0.63
11	3.6400	0.0614	-0.0602	-0.51	1.83

```
Average of cell averages          =        3.70022
Standard Deviation of cell averages  =     0.11770
Repeatability Standard Deviation   =       0.03363
Reproducibility Standard Deviation  =      0.12164
Critical Values h,k  =        2.34, 1.75
```

XRF Glass, Replicate 1

Material	Xbar	s_x	s_r	s_R	r	R
1	3.2355	0.0544	0.0129	0.0557	0.04	0.16
2	3.6822	0.0802	0.0323	0.0855	0.09	0.24

XRF Glass, Replicate 2

Material	Xbar	s_x	s_r	s_R	r	R
1	3.2339	0.0533	0.0138	0.0548	0.04	0.15
2	3.6691	0.0822	0.0857	0.1135	0.24	0.32

XRF Powder Replicate 1

Material	Xbar	s_x	s_r	s_R	r	R
1	3.1821	0.0872	0.0233	0.0898	0.07	0.25
2	3.7005	0.1161	0.0325	0.1198	0.09	0.34

XRF Powder Replicate 2

Material	Xbar	s_x	s_r	s_R	r	R
1	3.1702	0.1027	0.0231	0.1049	0.06	0.29
2	3.7002	0.1177	0.0336	0.1216	0.09	0.34

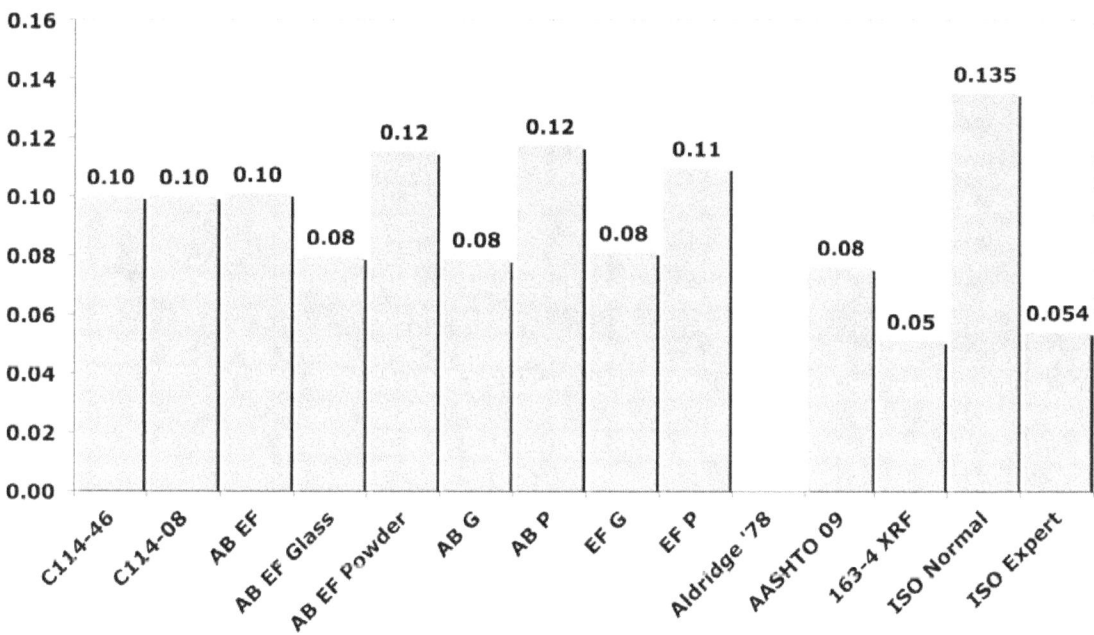

Figure 17 SO_3 precision statistics by method with bar chart comparing results to current and past ASTM C114 limits and previous studies on chemical analysis precision as 1σ, between lab (S_R).

Na₂O

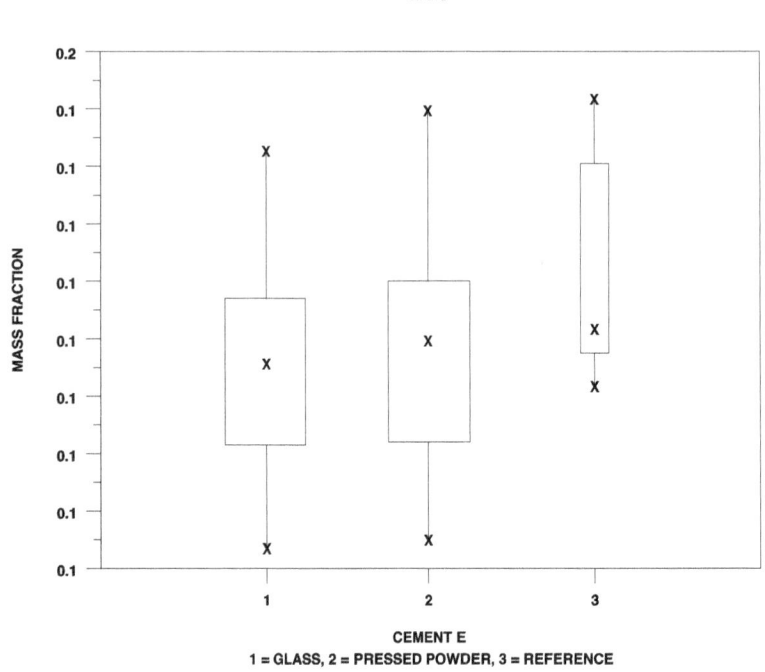

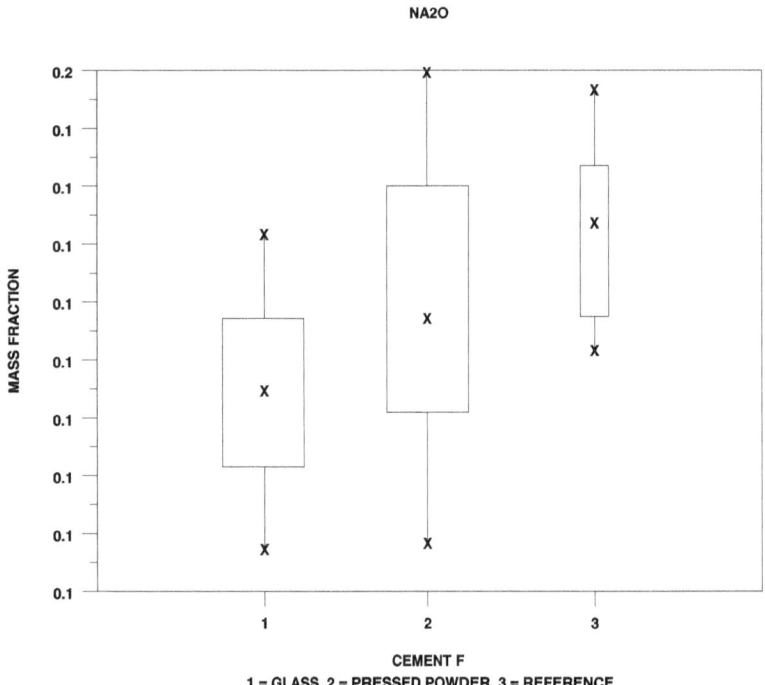

Figure 18 Box plots for Na₂O for XRF glass and powder, and reference methods.

Na₂O, Cements E, Glass, Rep. 1

Laboratory Number	Cell Mean	Cell SD	d	h	k
1	0.1580	0.0021	0.0035	0.22	0.30
2	0.1835	0.0083	0.0290	1.78	1.18
3	0.1350	0.0084	-0.0195	-1.20	1.19
4	0.1523	0.0025	-0.0022	-0.14	0.36
5	0.1299	0.0047	-0.0246	-1.51	0.67
6	0.1702	0.0023	0.0157	0.96	0.33
7	0.1410	0.0049	-0.0135	-0.83	0.71
8	0.1519	0.0080	-0.0026	-0.16	1.15
9	0.1600	0.0141	0.0055	0.34	2.02
10	0.1632	0.0050	0.0087	0.53	0.71

```
Average of cell averages          =        0.15448
Standard Deviation of cell averages  =     0.01628
Repeatability Standard Deviation   =       0.00700
Reproducibility Standard Deviation =       0.01749
Critical Values h,k  =        2.29, 1.74
```

Na₂O, Cements F, Glass, Rep. 1

Laboratory Number	Cell Mean	Cell SD	d	h	k
1	0.1503	0.0010	0.0056	0.39	0.17
2	0.1657	0.0059	0.0210	1.45	0.99
3	0.1283	0.0075	-0.0164	-1.13	1.27
4	0.1446	0.0015	-0.0001	-0.01	0.25
5	0.1211	0.0023	-0.0236	-1.63	0.39
6	0.1622	0.0037	0.0175	1.21	0.62
7	0.1338	0.0062	-0.0109	-0.75	1.05
8	0.1443	0.0124	-0.0004	-0.03	2.10
9	0.1400	0.0000	-0.0047	-0.32	0.00
10	0.1567	0.0067	0.0120	0.83	1.14

```
Average of cell averages          =        0.14470
Standard Deviation of cell averages  =     0.01447
Repeatability Standard Deviation   =       0.00594
Reproducibility Standard Deviation =       0.01545
Critical Values h,k  =        2.29, 1.74
```

Na₂O, Cements E, Glass, Rep. 2

Laboratory Number	Cell Mean	Cell SD	d	h	k
1	0.1593	0.0023	0.0041	0.28	0.47
2	0.1805	0.0065	0.0253	1.70	1.30
3	0.1433	0.0052	-0.0119	-0.80	1.03
4	0.1532	0.0014	-0.0020	-0.13	0.29
5	0.1298	0.0039	-0.0254	-1.71	0.79
6	0.1715	0.0031	0.0163	1.10	0.63
7	0.1425	0.0056	-0.0127	-0.86	1.13
8	0.1507	0.0087	-0.0045	-0.30	1.73
9	0.1567	0.0052	0.0015	0.10	1.03
10	0.1645	0.0039	0.0093	0.63	0.78

```
Average of cell averages            =       0.15521
Standard Deviation of cell averages =       0.01484
Repeatability Standard Deviation    =       0.00501
Reproducibility Standard Deviation  =       0.01553
Critical Values h,k  =       2.29, 1.74
```

Na₂O, Cements F, Glass, Rep. 1

Laboratory Number	Cell Mean	Cell SD	d	h	k
1	0.1503	0.0008	0.0036	0.27	0.14
2	0.1682	0.0051	0.0214	1.59	0.85
3	0.1350	0.0055	-0.0117	-0.87	0.92
4	0.1441	0.0016	-0.0026	-0.20	0.27
5	0.1235	0.0027	-0.0233	-1.72	0.45
6	0.1615	0.0019	0.0148	1.09	0.31
7	0.1347	0.0062	-0.0121	-0.89	1.04
8	0.1505	0.0140	0.0038	0.28	2.35
9	0.1433	0.0052	-0.0034	-0.25	0.87
10	0.1563	0.0049	0.0096	0.71	0.82

```
Average of cell averages            =       0.14674
Standard Deviation of cell averages =       0.01350
Repeatability Standard Deviation    =       0.00594
Reproducibility Standard Deviation  =       0.01455
Critical Values h,k  =       2.29, 1.74
```

Na$_2$O, Cements E, Powder, Rep. 1

Laboratory Number	Cell Mean	Cell SD	d	h	k
1	0.1600	0.0025	0.0010	0.05	0.36
2	0.1433	0.0052	-0.0156	-0.78	0.72
3	0.1400	0.0110	-0.0190	-0.94	1.54
4	0.1633	0.0052	0.0044	0.22	0.72
5	0.1567	0.0052	-0.0023	-0.11	0.72
6	0.1767	0.0151	0.0177	0.88	2.11
7	0.1495	0.0036	-0.0095	-0.47	0.51
8	0.1750	0.0055	0.0160	0.80	0.77
9	0.1285	0.0022	-0.0305	-1.52	0.31
10	0.1967	0.0052	0.0377	1.87	0.72

```
Average of cell averages          =        0.15896
Standard Deviation of cell averages =      0.02011
Repeatability Standard Deviation  =        0.00713
Reproducibility Standard Deviation =       0.02114
Critical Values h,k  =          2.29, 1.74
```

Na$_2$O, Cements F, Powder, Rep. 1

Laboratory Number	Cell Mean	Cell SD	d	h	k
1	0.1570	0.0028	-0.0009	-0.04	0.46
2	0.1400	0.0000	-0.0179	-0.82	0.00
3	0.1367	0.0151	-0.0213	-0.97	2.54
4	0.1700	0.0000	0.0121	0.55	0.00
5	0.1550	0.0055	-0.0029	-0.13	0.92
6	0.1767	0.0052	0.0187	0.86	0.87
7	0.1455	0.0027	-0.0124	-0.57	0.45
8	0.1817	0.0041	0.0237	1.08	0.69
9	0.1245	0.0044	-0.0334	-1.53	0.74
10	0.1925	0.0042	0.0345	1.58	0.71

```
Average of cell averages          =        0.15795
Standard Deviation of cell averages =      0.02188
Repeatability Standard Deviation  =        0.00593
Reproducibility Standard Deviation =       0.02254
Critical Values h,k  =          2.29, 1.74
```

Na₂O, Cements E, Powder, Rep. 2

Laboratory Number	Cell Mean	Cell SD	d	h	k
1	0.1623	0.0023	0.0025	0.13	0.29
2	0.1417	0.0041	-0.0182	-0.92	0.53
3	0.1400	0.0141	-0.0198	-1.00	1.83
4	0.1650	0.0055	0.0052	0.26	0.71
5	0.1600	0.0000	0.0002	0.01	0.00
6	0.1817	0.0172	0.0218	1.11	2.22
7	0.1532	0.0025	-0.0067	-0.34	0.32
8	0.1767	0.0052	0.0168	0.85	0.67
9	0.1277	0.0043	-0.0321	-1.63	0.56
10	0.1900	0.0000	0.0302	1.53	0.00

```
Average of cell averages          =          0.15982
Standard Deviation of cell averages  =          0.01973
Repeatability Standard Deviation   =          0.00774
Reproducibility Standard Deviation  =          0.02096
Critical Values h,k  =        2.29, 1.74
```

Na₂O, Cements F, Powder, Rep. 2

Laboratory Number	Cell Mean	Cell SD	d	h	k
1	0.1543	0.0054	-0.0028	-0.13	0.77
2	0.1383	0.0041	-0.0188	-0.90	0.58
3	0.1383	0.0133	-0.0188	-0.90	1.90
4	0.1650	0.0055	0.0079	0.38	0.78
5	0.1533	0.0052	-0.0038	-0.18	0.74
6	0.1733	0.0103	0.0162	0.78	1.48
7	0.1488	0.0039	-0.0083	-0.40	0.55
8	0.1800	0.0063	0.0229	1.10	0.91
9	0.1260	0.0046	-0.0311	-1.49	0.66
10	0.1933	0.0052	0.0362	1.74	0.74

```
Average of cell averages          =          0.15709
Standard Deviation of cell averages  =          0.02088
Repeatability Standard Deviation   =          0.00699
Reproducibility Standard Deviation  =          0.02183
Critical Values h,k  =        2.29, 1.74
```

XRF Glass, Replicate 1

Material	Xbar	s_x	s_r	s_R	r	R
1	0.1545	0.0163	0.0070	0.0175	0.02	0.05
2	0.1447	0.0145	0.0059	0.0155	0.02	0.04

XRF Glass, Replicate 2

Material	Xbar	s_x	s_r	s_R	r	R
1	0.1552	0.0148	0.0050	0.0155	0.01	0.04
2	0.1467	0.0135	0.0059	0.0145	0.02	0.04

XRF Powder, Replicate 1

Material	Xbar	s_x	s_r	s_R	r	R
1	0.1590	0.0201	0.0071	0.0211	0.02	0.06
2	0.1579	0.0219	0.0059	0.0225	0.02	0.06

XRF Powder, Replicate 2

Material	Xbar	s_x	s_r	s_R	r	R
1	0.1598	0.0197	0.0077	0.0210	0.02	0.06
2	0.1571	0.0209	0.0070	0.0218	0.02	0.06

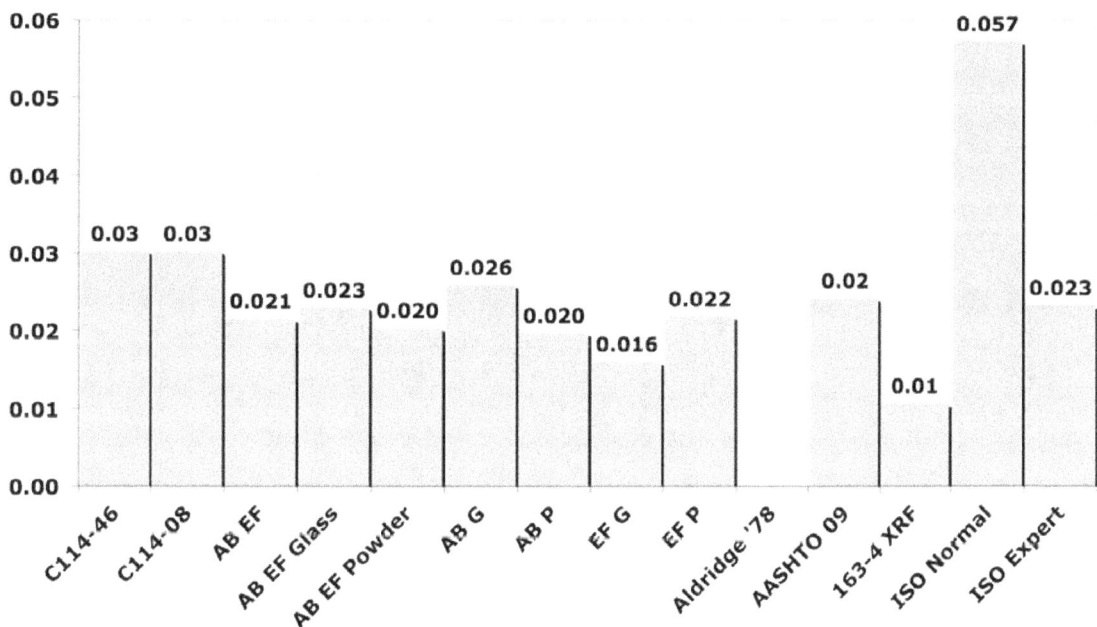

Figure 19 Na_2O precision statistics by method with bar chart comparing results to current and past ASTM C114 limits and previous studies on chemical analysis precision as 1σ, between lab (S_R).

55

K₂O

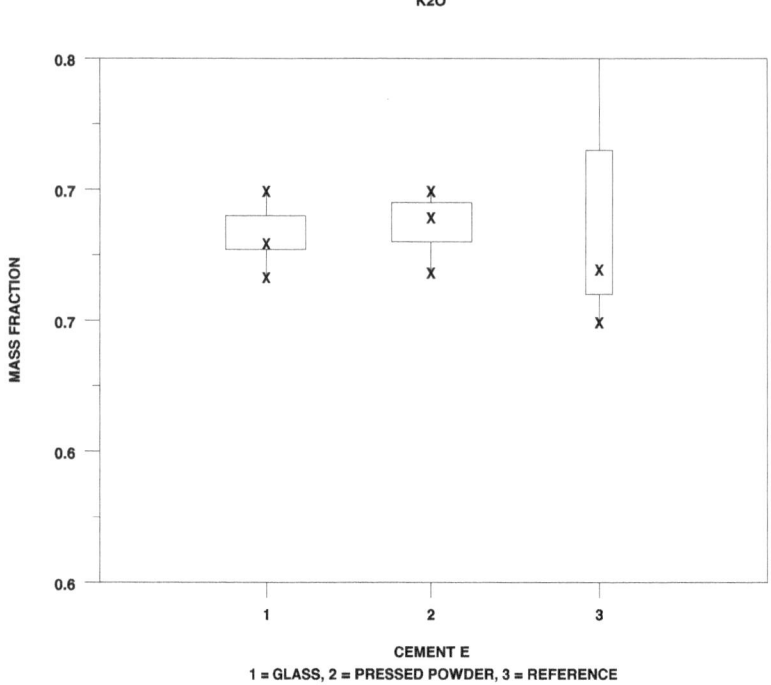

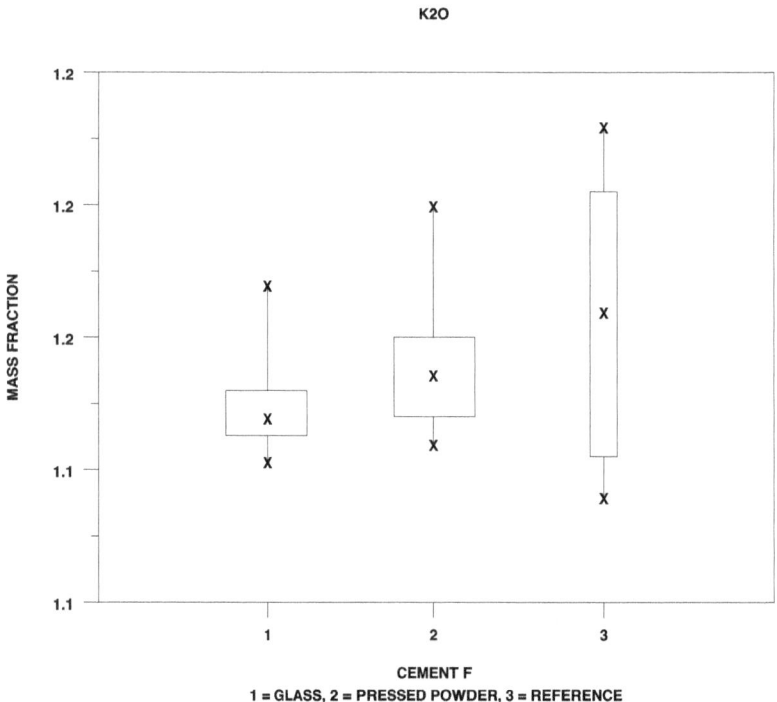

Figure 20 Box plots for K₂O for XRF glass and powder, and reference methods.

K$_2$O, Cements E, Glass, Rep. 1

Laboratory Number	Cell Mean	Cell SD	d	h	k
1	0.7238	0.0044	-0.0087	-1.06	1.52
2	0.7383	0.0041	0.0058	0.70	1.43
3	0.7364	0.0030	0.0039	0.47	1.06
4	0.7256	0.0032	-0.0070	-0.85	1.14
5	0.7317	0.0041	-0.0009	-0.11	1.43
6	0.7400	0.0000	0.0075	0.91	0.00
7	0.7277	0.0027	-0.0049	-0.59	0.93
8	0.7326	0.0016	0.0001	0.01	0.55
9	0.7300	0.0000	-0.0025	-0.31	0.00
10	0.7218	0.0029	-0.0107	-1.30	1.00
11	0.7500	0.0000	0.0175	2.12	0.00

```
Average of cell averages          =        0.73254
Standard Deviation of cell averages  =     0.00824
Repeatability Standard Deviation  =        0.00286
Reproducibility Standard Deviation  =      0.00864
Critical Values h,k  =        2.34, 1.75
```

K$_2$O, Cements F, Glass, Rep. 1

Laboratory Number	Cell Mean	Cell SD	d	h	k
1	1.1612	0.0066	-0.0130	-0.84	1.36
2	1.1850	0.0055	0.0108	0.69	1.12
3	1.1794	0.0032	0.0052	0.33	0.65
4	1.1636	0.0056	-0.0106	-0.68	1.14
5	1.1817	0.0041	0.0075	0.48	0.83
6	1.1733	0.0052	-0.0009	-0.06	1.06
7	1.1665	0.0029	-0.0077	-0.49	0.59
8	1.1620	0.0047	0.0122	-0.78	0.96
9	1.1700	0.0063	-0.0042	-0.27	1.29
10	1.1602	0.0028	-0.0140	-0.90	0.57
11	1.2133	0.0052	0.0391	2.51	1.06

```
Average of cell averages          =        1.17420
Standard Deviation of cell averages  =     0.01561
Repeatability Standard Deviation  =        0.00489
Reproducibility Standard Deviation  =      0.01623
Critical Values h,k  =        2.34, 1.75
```

K$_2$O, Cements E, Glass, Rep. 2

Laboratory Number	Cell Mean	Cell SD	d	h	k
1	0.7238	0.0033	-0.0071	-1.16	0.90
2	0.7350	0.0055	0.0040	0.65	1.49
3	0.7354	0.0026	0.0045	0.73	0.72
4	0.7263	0.0048	-0.0046	-0.75	1.30
5	0.7300	0.0000	-0.0010	-0.16	0.00
6	0.7350	0.0055	0.0040	0.65	1.49
7	0.7273	0.0027	-0.0036	-0.59	0.73
8	0.7326	0.0020	0.0016	0.27	0.54
9	0.7300	0.0000	-0.0010	-0.16	0.00
10	0.7218	0.0031	-0.0091	-1.48	0.84
11	0.7433	0.0052	0.0124	2.01	1.41

```
Average of cell averages           =       0.73098
Standard Deviation of cell averages =      0.00616
Repeatability Standard Deviation   =       0.00366
Reproducibility Standard Deviation =       0.00701
Critical Values h,k  =        2.34, 1.75
```

K$_2$O, Cements F, Glass, Rep. 2

Laboratory Number	Cell Mean	Cell SD	d	h	k
1	1.1615	0.0072	-0.0138	-0.85	1.54
2	1.1850	0.0055	0.0097	0.60	1.17
3	1.1786	0.0013	0.0033	0.20	0.27
4	1.1634	0.0032	-0.0119	-0.73	0.69
5	1.1867	0.0052	0.0114	0.70	1.10
6	1.1733	0.0052	-0.0020	-0.12	1.10
7	1.1708	0.0042	-0.0045	-0.27	0.90
8	1.1620	0.0043	-0.0133	-0.82	0.92
9	1.1733	0.0052	-0.0020	-0.12	1.10
10	1.1585	0.0014	-0.0168	-1.04	0.29
11	1.2150	0.0055	0.0397	2.45	1.17

```
Average of cell averages           =       1.17529
Standard Deviation of cell averages =      0.01621
Repeatability Standard Deviation   =       0.00469
Reproducibility Standard Deviation =       0.01677
Critical Values h,k  =        2.34, 1.75
```

K$_2$O, Cements E, Powder, Rep. 1

Laboratory Number	Cell Mean	Cell SD	d	h	k
1	0.7417	0.0075	0.0044	0.53	1.43
2	0.7333	0.0052	-0.0039	-0.47	0.98
3	0.7383	0.0041	0.0011	0.13	0.77
4	0.7483	0.0041	0.0111	1.34	0.77
5	0.7367	0.0052	-0.0006	-0.07	0.98
6	0.7467	0.0052	0.0094	1.14	0.98
7	0.7287	0.0037	-0.0086	-1.04	0.69
8	0.7300	0.0000	-0.0072	-0.88	0.00
9	0.7450	0.0084	0.0078	0.94	1.58
10	0.7237	0.0051	-0.0135	-1.63	0.96

```
Average of cell averages            =        0.73724
Standard Deviation of cell averages =        0.00828
Repeatability Standard Deviation    =        0.00528
Reproducibility Standard Deviation  =        0.00958
Critical Values h,k  =        2.29, 1.74
```

K$_2$O, Cements F, Powder, Rep. 1

Laboratory Number	Cell Mean	Cell SD	d	h	k
1	1.1817	0.0117	-0.0080	-0.37	1.49
2	1.1800	0.0089	-0.0096	-0.44	1.14
3	1.1883	0.0075	-0.0013	-0.06	0.96
4	1.1983	0.0075	0.0087	0.40	0.96
5	1.1883	0.0075	-0.0013	-0.06	0.96
6	1.2383	0.0075	0.0487	2.25	0.96
7	1.1683	0.0055	-0.0213	-0.98	0.70
8	1.1650	0.0055	-0.0246	-1.14	0.70
9	1.2100	0.0089	0.0204	0.94	1.14
10	1.1780	0.0055	-0.0117	-0.54	0.70

```
Average of cell averages            =        1.18963
Standard Deviation of cell averages =        0.02165
Repeatability Standard Deviation    =        0.00783
Reproducibility Standard Deviation  =        0.02280
Critical Values h,k  =        2.29, 1.74
```

K$_2$O, Cements E, Powder, Rep. 2

Laboratory Number	Cell Mean	Cell SD	d	h	k
1	0.7383	0.0041	0.0014	0.17	0.82
2	0.7317	0.0041	-0.0053	-0.66	0.82
3	0.7400	0.0063	0.0031	0.38	1.26
4	0.7467	0.0052	0.0097	1.22	1.03
5	0.7383	0.0041	0.0014	0.17	0.82
6	0.7450	0.0055	0.0081	1.01	1.09
7	0.7285	0.0036	-0.0084	-1.05	0.71
8	0.7300	0.0000	-0.0069	-0.87	0.00
9	0.7467	0.0082	0.0097	1.22	1.63
10	0.7242	0.0049	-0.0127	-1.59	0.99

```
Average of cell averages            =        0.73694
Standard Deviation of cell averages =        0.00800
Repeatability Standard Deviation    =        0.00500
Reproducibility Standard Deviation  =        0.00922
Critical Values h,k  =         2.29, 1.74
```

K$_2$O, CementsF, Powder, Rep. 2

Laboratory Number	Cell Mean	Cell SD	d	h	k
1	1.1833	0.0082	-0.0062	-0.28	0.85
2	1.1800	0.0110	-0.0095	-0.44	1.14
3	1.1900	0.0063	0.0005	0.02	0.66
4	1.2017	0.0098	0.0122	0.56	1.02
5	1.1883	0.0075	-0.0012	-0.05	0.78
6	1.2367	0.0052	0.0472	2.17	0.54
7	1.1698	0.0055	-0.0197	-0.90	0.57
8	1.1667	0.0052	-0.0228	-1.05	0.54
9	1.2100	0.0063	0.0205	0.94	0.66
10	1.1686	0.0205	-0.0209	-0.96	2.13

```
Average of cell averages            =        1.18951
Standard Deviation of cell averages =        0.02178
Repeatability Standard Deviation    =        0.00960
Reproducibility Standard Deviation  =        0.02348
Critical Values h,k  =         2.29, 1.74
```

XRF Glass, Replicate 1

Material	Xbar	s_x	s_r	s_R	r	R
1	0.7325	0.0082	0.0029	0.0086	0.01	0.02
2	1.1742	0.0156	0.0049	0.0162	0.01	0.05

XRF Glass, Replicate 2

Material	Xbar	s_x	s_r	s_R	r	R
1	0.7310	0.0062	0.0037	0.0070	0.01	0.02
2	1.1753	0.0162	0.0047	0.0168	0.01	0.05

XRF Powder Replicate 1

Material	Xbar	s_x	s_r	s_R	r	R
1	0.7372	0.0083	0.0053	0.0096	0.01	0.03
2	1.1896	0.0217	0.0078	0.0228	0.02	0.06

XRF Powder Replicate 2

Material	Xbar	s_x	s_r	s_R	r	R
1	0.7369	0.0080	0.0050	0.0092	0.01	0.03
2	1.1895	0.0218	0.0096	0.0235	0.03	0.07

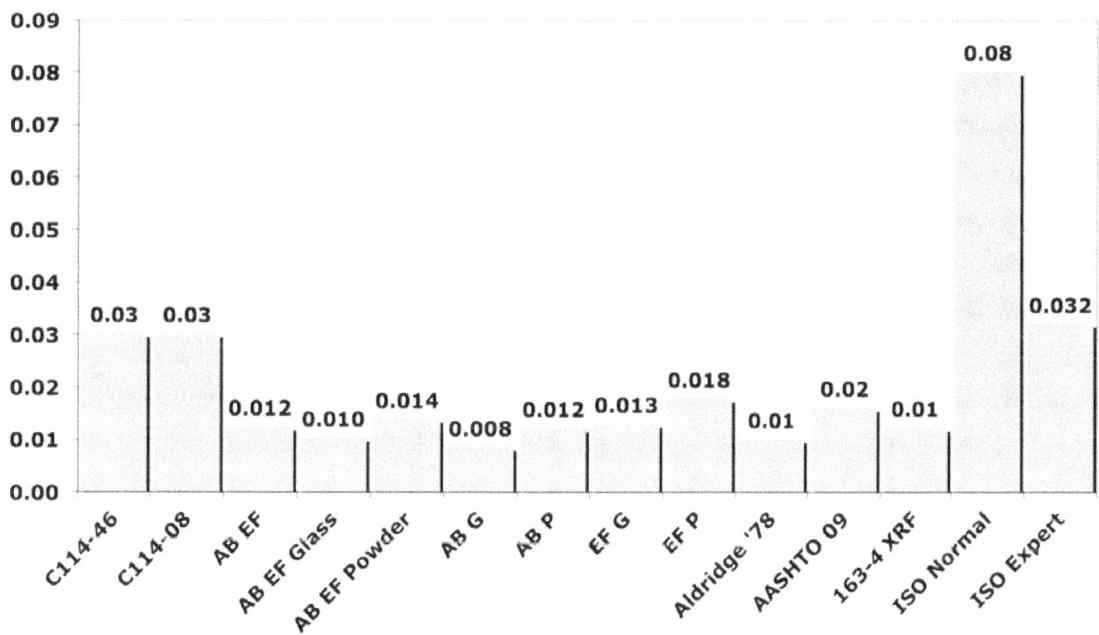

Figure 21 K_2O precision statistics by method with bar chart comparing results to current and past ASTM C114 limits and previous studies on chemical analysis precision as 1σ, between lab (S_R).

61

TiO₂

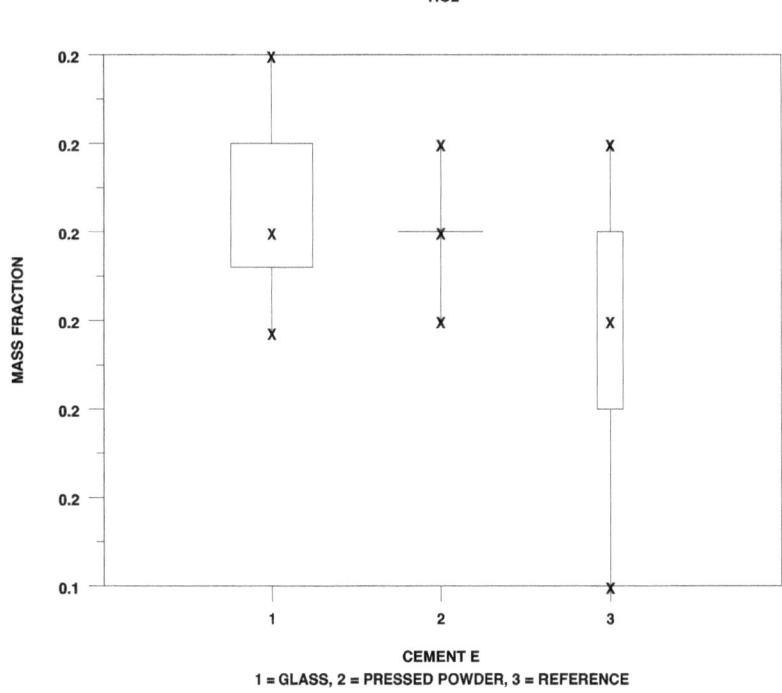

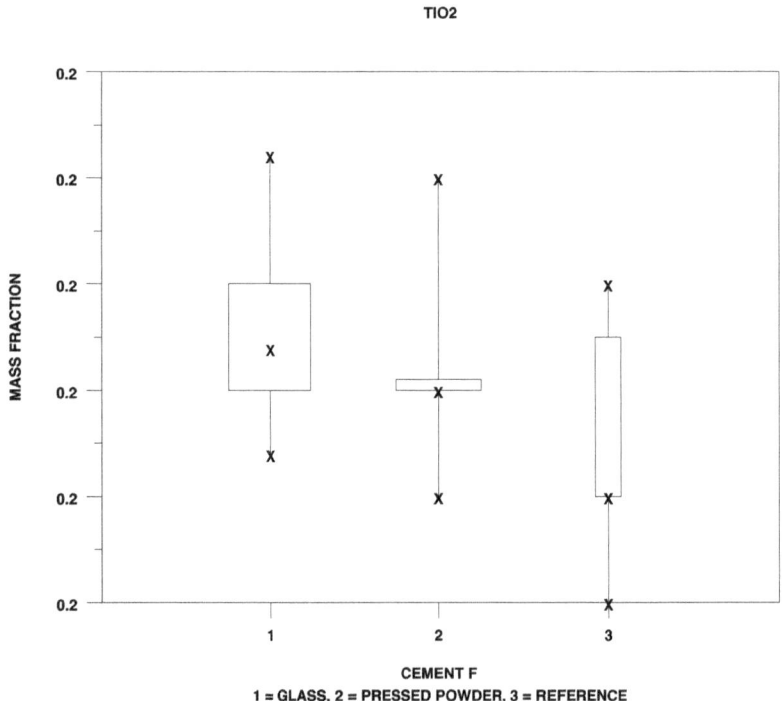

Figure 22 Box plots for TiO₂ for XRF glass and powder, and reference methods.

TiO₂ Cements E, Glass, Rep. 1

Laboratory Number	Cell Mean	Cell SD	d	h	k
1	0.2280	0.0030	-0.0035	-0.44	0.76
2	0.2323	0.0050	0.0009	0.11	1.29
3	0.2350	0.0055	0.0035	0.46	1.40
4	0.2302	0.0006	-0.0012	-0.16	0.14
5	0.2469	0.0017	0.0155	1.99	0.43
6	0.2283	0.0041	-0.0031	-0.40	1.05
7	0.2217	0.0041	-0.0098	-1.26	1.05
8	0.2247	0.0014	-0.0068	-0.87	0.35
9	0.2220	0.0032	-0.0095	-1.22	0.82
10	0.2350	0.0055	0.0035	0.46	1.40
11	0.2300	0.0037	-0.0015	-0.19	0.94
12	0.2433	0.0052	0.0119	1.53	1.32

```
Average of cell averages              =        0.23145
Standard Deviation of cell averages   =        0.00777
Repeatability Standard Deviation      =        0.00391
Reproducibility Standard Deviation    =        0.00855
Critical Values h,k  =        2.38, 1.76
```

TiO₂ Cements F, Glass, Rep. 1

Laboratory Number	Cell Mean	Cell SD	d	h	k
1	0.2243	0.0026	-0.0019	-0.22	0.80
2	0.2188	0.0038	-0.0073	-0.89	1.17
3	0.2317	0.0041	0.0055	0.66	1.26
4	0.2246	0.0008	-0.0016	-0.19	0.24
5	0.2396	0.0019	0.0134	1.62	0.57
6	0.2217	0.0041	-0.0045	-0.54	1.26
7	0.2200	0.0000	-0.0062	-0.75	0.00
8	0.2190	0.0021	-0.0072	-0.87	0.65
9	0.2162	0.0024	-0.0100	-1.21	0.74
10	0.2350	0.0055	0.0088	1.06	1.70
11	0.2234	0.0052	-0.0028	-0.34	1.60
12	0.2400	0.0000	0.0138	1.67	0.00

Average of cell averages   =   0.22618
Standard Deviation of cell averages =   0.00829
Repeatability Standard Deviation   =   0.00323
Reproducibility Standard Deviation =   0.00880
Critical Values h,k =   2.38, 1.76

TiO$_2$ Cements E, Glass, Rep. 2

Laboratory Number	Cell Mean	Cell SD	d	h	k
1	0.2277	0.0024	-0.0037	-0.50	0.61
2	0.2275	0.0061	-0.0039	-0.52	1.53
3	0.2383	0.0041	0.0070	0.94	1.03
4	0.2292	0.0012	-0.0021	-0.28	0.30
5	0.2418	0.0036	0.0104	1.41	0.90
6	0.2300	0.0000	-0.0014	-0.18	0.00
7	0.2217	0.0041	-0.0097	-1.31	1.03
8	0.2260	0.0021	-0.0054	-0.72	0.53
9	0.2217	0.0013	-0.0096	-1.30	0.34
10	0.2333	0.0052	0.0020	0.27	1.30
11	0.2341	0.0060	0.0027	0.36	1.50
12	0.2450	0.0055	0.0136	1.84	1.38

```
Average of cell averages            =        0.23136
Standard Deviation of cell averages =        0.00742
Repeatability Standard Deviation    =        0.00396
Reproducibility Standard Deviation  =        0.00825
Critical Values h,k  =        2.38, 1.76
```

TiO$_2$ Cements F, Glass, Rep. 2

Laboratory Number	Cell Mean	Cell SD	d	h	k
1	0.2242	0.0021	-0.0024	-0.31	0.63
2	0.2237	0.0029	-0.0029	-0.38	0.84
3	0.2267	0.0052	0.0001	0.02	1.51
4	0.2247	0.0009	-0.0018	-0.24	0.27
5	0.2377	0.0042	0.0112	1.49	1.22
6	0.2250	0.0055	-0.0015	-0.20	1.60
7	0.2200	0.0000	-0.0065	-0.87	0.00
8	0.2190	0.0017	-0.0075	-1.00	0.49
9	0.2164	0.0013	-0.0102	-1.35	0.37
10	0.2333	0.0052	0.0068	0.90	1.51
11	0.2259	0.0022	-0.0006	-0.08	0.64
12	0.2417	0.0041	0.0151	2.01	1.19

```
Average of cell averages            =        0.22652
Standard Deviation of cell averages =        0.00753
Repeatability Standard Deviation    =        0.00342
Reproducibility Standard Deviation  =        0.00815
Critical Values h,k  =        2.38, 1.76
```

TiO₂ Cements E, Powder, Rep. 1

Laboratory Number	Cell Mean	Cell SD	*d*	*h*	*k*
1	0.2283	0.0041	-0.0017	-0.44	1.25
2	0.2300	0.0000	-0.0001	-0.02	0.00
3	0.2400	0.0000	0.0099	2.49	0.00
4	0.2300	0.0000	-0.0001	-0.02	0.00
5	0.2250	0.0055	-0.0051	-1.27	1.68
6	0.2300	0.0000	-0.0001	-0.02	0.00
7	0.2300	0.0000	-0.0001	-0.02	0.00
8	0.2250	0.0055	-0.0051	-1.27	1.68
9	0.2305	0.0019	0.0004	0.10	0.57
10	0.2333	0.0052	0.0033	0.82	1.58
11	0.2272	0.0038	-0.0029	-0.74	1.18
12	0.2317	0.0026	0.0016	0.40	0.79

```
Average of cell averages            =        0.23008
Standard Deviation of cell averages =        0.00399
Repeatability Standard Deviation    =        0.00327
Reproducibility Standard Deviation  =        0.00498
Critical Values h,k  =         2.38, 1.76
```

TiO₂ Cements F, Powder, Rep. 1

Laboratory Number	Cell Mean	Cell SD	*d*	*h*	*k*
1	0.2233	0.0052	0.0018	0.30	1.19
2	0.2217	0.0041	0.0001	0.02	0.94
3	0.2367	0.0052	0.0151	2.52	1.19
4	0.2217	0.0041	0.0001	0.02	0.94
5	0.2217	0.0075	0.0001	0.02	1.74
6	0.2250	0.0055	0.0035	0.58	1.26
7	0.2200	0.0000	-0.0015	-0.26	0.00
8	0.2100	0.0000	-0.0115	-1.92	0.00
9	0.2197	0.0012	-0.0019	-0.31	0.28
10	0.2200	0.0063	-0.0015	-0.26	1.46
11	0.2187	0.0033	-0.0028	-0.47	0.77
12	0.2200	0.0000	-0.0015	-0.26	0.00

```
Average of cell averages            =        0.22153
Standard Deviation of cell averages =        0.00600
Repeatability Standard Deviation    =        0.00434
Reproducibility Standard Deviation  =        0.00719
Critical Values h,k  =         2.38, 1.76
```

TiO$_2$ Cements E, Powder, Rep. 2

Laboratory Number	Cell Mean	Cell SD	d	h	k
1	0.2300	0.0000	0.0000	0.01	0.00
2	0.2300	0.0000	0.0000	0.01	0.00
3	0.2400	0.0000	0.0100	2.28	0.00
4	0.2300	0.0000	0.0000	0.01	0.00
5	0.2233	0.0052	-0.0066	-1.50	1.67
6	0.2317	0.0041	0.0017	0.39	1.32
7	0.2300	0.0000	0.0000	0.01	0.00
8	0.2250	0.0055	-0.0050	-1.12	1.77
9	0.2282	0.0015	-0.0018	-0.41	0.48
10	0.2350	0.0055	0.0050	1.14	1.77
11	0.2263	0.0031	-0.0037	-0.83	1.00
12	0.2300	0.0000	0.0000	0.01	0.00

```
Average of cell averages              =        0.22996
Standard Deviation of cell averages   =        0.00441
Repeatability Standard Deviation      =        0.00310
Reproducibility Standard Deviation    =        0.00524
Critical Values h,k  =        2.38, 1.76
```

TiO$_2$ Cements F, Powder, Rep. 2

Laboratory Number	Cell Mean	Cell SD	d	h	k
1	0.2217	0.0041	0.0004	0.07	1.04
2	0.2200	0.0000	-0.0013	-0.21	0.00
3	0.2383	0.0041	0.0171	2.85	1.04
4	0.2217	0.0041	0.0004	0.07	1.04
5	0.2183	0.0041	-0.0029	-0.49	1.04
6	0.2233	0.0052	0.0021	0.35	1.31
7	0.2200	0.0000	-0.0013	-0.21	0.00
8	0.2133	0.0052	-0.0079	-1.33	1.31
9	0.2200	0.0017	-0.0013	-0.21	0.42
10	0.2217	0.0075	0.0004	0.07	1.91
11	0.2169	0.0026	-0.0044	-0.74	0.65
12	0.2200	0.0000	-0.0013	-0.21	0.00

```
Average of cell averages              =        0.22127
Standard Deviation of cell averages   =        0.00598
Repeatability Standard Deviation      =        0.00394
Reproducibility Standard Deviation    =        0.00698
Critical Values h,k  =        2.38, 1.76
```

XRF Glass, Replicate 1

Material	Xbar	s_x	s_r	s_R	r	R
1	0.2315	0.0078	0.0039	0.0085	0.01	0.02
2	0.2262	0.0083	0.0032	0.0088	0.01	0.02

XRF Glass, Replicate 2

Material	Xbar	s_x	s_r	s_R	r	R
1	0.2314	0.0074	0.0040	0.0083	0.01	0.02
2	0.2265	0.0075	0.0034	0.0082	0.01	0.02

XRF Powder, Replicate 1

Material	Xbar	s_x	s_r	s_R	r	R
1	0.2301	0.0040	0.0033	0.0050	0.01	0.01
2	0.2215	0.0060	0.0043	0.0072	0.01	0.02

XRF Powder, Replicate 2

Material	Xbar	s_x	s_r	s_R	r	R
1	0.2300	0.0044	0.0031	0.0052	0.01	0.01
2	0.2213	0.0060	0.0039	0.0070	0.01	0.02

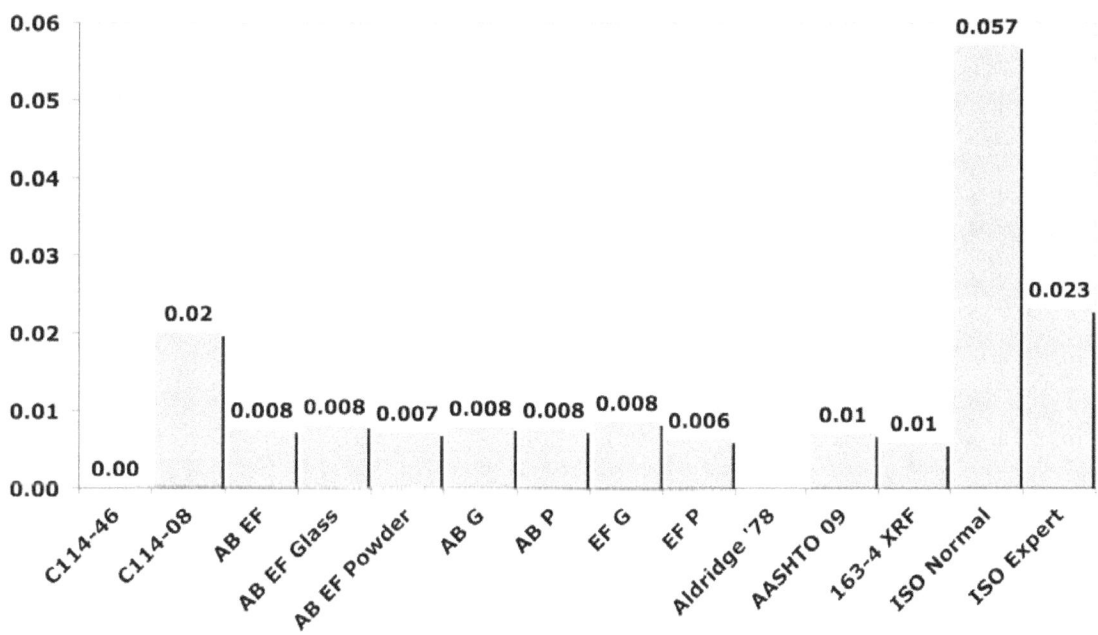

Figure 23 TiO$_2$ precision statistics by method with bar chart comparing results to current and past ASTM C114 limits and previous studies on chemical analysis precision as 1σ, between lab (S$_R$).

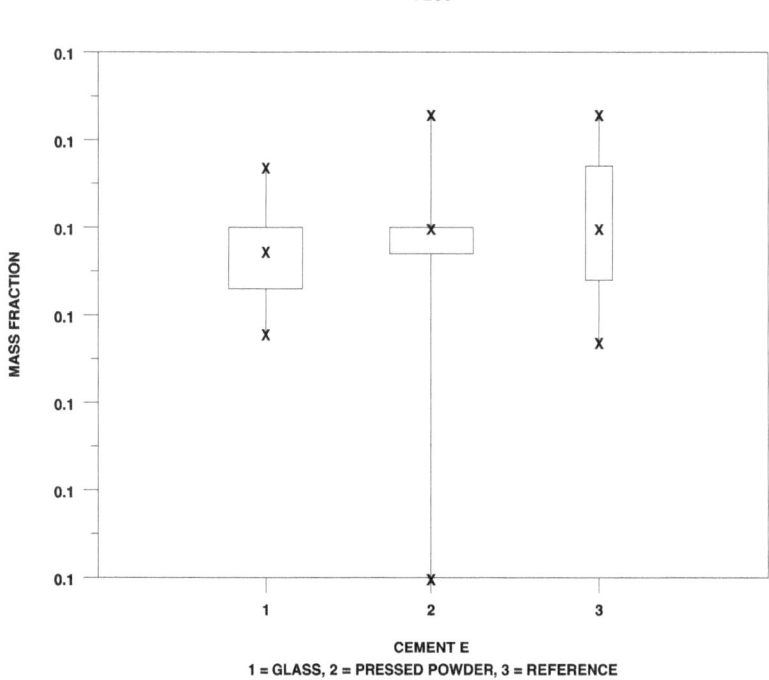

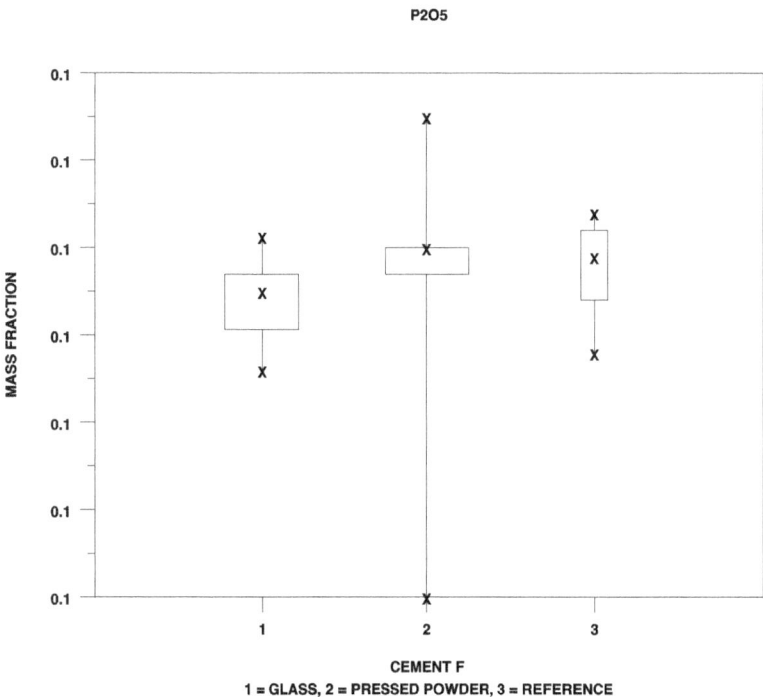

Figure 24 Box plots for P$_2$O$_5$ for XRF glass and powder, and reference methods.

P$_2$O$_5$, Cements E, Glass, Rep. 1

Laboratory Number	Cell Mean	Cell SD	d	h	k
1	0.1407	0.0031	0.0044	1.16	1.52
2	0.1308	0.0017	-0.0055	-1.44	0.84
3	0.1369	0.0010	0.0006	0.17	0.46
4	0.1353	0.0025	-0.0010	-0.26	1.23
5	0.1400	0.0000	0.0037	0.98	0.00
6	0.1367	0.0008	0.0004	0.10	0.40
7	0.1395	0.0016	0.0032	0.85	0.80
8	0.1300	0.0000	-0.0063	-1.66	0.00
9	0.1367	0.0038	0.0004	0.12	1.86

```
Average of cell averages          =        0.13629
Standard Deviation of cell averages =      0.00378
Repeatability Standard Deviation  =        0.00206
Reproducibility Standard Deviation =       0.00422
Critical Values h,k  =        2.23, 1.73
```

P$_2$O$_5$, Cements F, Glass, Rep. 1

Laboratory Number	Cell Mean	Cell SD	d	h	k
1	0.1367	0.0010	0.0022	0.61	0.51
2	0.1300	0.0030	-0.0045	-1.27	1.45
3	0.1340	0.0010	-0.0005	-0.14	0.49
4	0.1311	0.0028	-0.0034	-0.96	1.38
5	0.1400	0.0000	0.0055	1.55	0.00
6	0.1347	0.0005	0.0002	0.05	0.25
7	0.1370	0.0028	0.0025	0.71	1.35
8	0.1300	0.0000	-0.0045	-1.27	0.00
9	0.1371	0.0033	0.0026	0.73	1.61

```
Average of cell averages          =        0.13450
Standard Deviation of cell averages =      0.00354
Repeatability Standard Deviation  =        0.00204
Reproducibility Standard Deviation =       0.00400
Critical Values h,k  =        2.23, 1.73
```

P$_2$O$_5$, Cements E, Glass, Rep. 2

Laboratory Number	Cell Mean	Cell SD	d	h	k
1	0.1397	0.0010	0.0029	0.71	0.59
2	0.1302	0.0015	-0.0066	-1.64	0.83
3	0.1368	0.0009	0.0000	-0.01	0.49
4	0.1373	0.0024	0.0005	0.13	1.35
5	0.1400	0.0000	0.0032	0.79	0.00
6	0.1372	0.0008	0.0004	0.09	0.43
7	0.1393	0.0019	0.0025	0.62	1.05
8	0.1300	0.0000	-0.0068	-1.68	0.00
9	0.1408	0.0038	0.0040	0.99	2.15

```
Average of cell averages          =        0.13681
Standard Deviation of cell averages  =     0.00406
Repeatability Standard Deviation   =       0.00177
Reproducibility Standard Deviation  =      0.00436
Critical Values h,k   =        2.23, 1.73
```

P$_2$O$_5$, Cements F, Glass, Rep. 1

Laboratory Number	Cell Mean	Cell SD	d	h	k
1	0.1372	0.0010	0.0021	0.69	0.52
2	0.1318	0.0034	-0.0032	-1.06	1.83
3	0.1344	0.0006	-0.0007	-0.24	0.32
4	0.1335	0.0025	-0.0015	-0.50	1.31
5	0.1400	0.0000	0.0049	1.62	0.00
6	0.1350	0.0013	-0.0001	-0.02	0.68
7	0.1370	0.0025	0.0019	0.63	1.35
8	0.1300	0.0000	-0.0051	-1.66	0.00
9	0.1367	0.0021	0.0017	0.55	1.13

```
Average of cell averages          =        0.13507
Standard Deviation of cell averages  =     0.00305
Repeatability Standard Deviation   =       0.00187
Reproducibility Standard Deviation  =      0.00349
Critical Values h,k   =        2.23, 1.73
```

P₂O₅, Cements E, Powder, Rep. 1

Laboratory Number	Cell Mean	Cell SD	d	h	k
1	0.1373	0.0014	0.0005	0.04	0.48
2	0.1400	0.0000	0.0032	0.25	0.00
3	0.1333	0.0052	-0.0035	-0.26	1.82
4	0.1450	0.0055	0.0082	0.63	1.94
5	0.1400	0.0000	0.0032	0.25	0.00
6	0.1380	0.0015	0.0012	0.09	0.55
7	0.1400	0.0000	0.0032	0.25	0.00
8	0.1000	0.0000	-0.0368	-2.81	0.00
9	0.1517	0.0010	0.0149	1.14	0.36
10	0.1375	0.0014	0.0007	0.05	0.49
11	0.1418	0.0049	0.0050	0.38	1.74

```
Average of cell averages              =        0.13679
Standard Deviation of cell averages   =        0.01309
Repeatability Standard Deviation      =        0.00283
Reproducibility Standard Deviation    =        0.01334
Critical Values h,k   =          2.34, 1.75
```

P₂O₅, Cements F, Powder, Rep. 1

Laboratory Number	Cell Mean	Cell SD	d	h	k
1	0.1363	0.0012	-0.0011	-0.10	0.37
2	0.1400	0.0000	0.0025	0.22	0.00
3	0.1350	0.0055	-0.0025	-0.22	1.67
4	0.1417	0.0041	0.0042	0.37	1.24
5	0.1383	0.0041	0.0009	0.08	1.24
6	0.1405	0.0044	0.0030	0.27	1.34
7	0.1400	0.0000	0.0025	0.22	0.00
8	0.1067	0.0052	-0.0308	-2.72	1.57
9	0.1538	0.0010	0.0164	1.44	0.30
10	0.1417	0.0015	0.0042	0.37	0.46
11	0.1382	0.0022	0.0008	0.07	0.66

```
Average of cell averages              =        0.13748
Standard Deviation of cell averages   =        0.01132
Repeatability Standard Deviation      =        0.00329
Reproducibility Standard Deviation    =        0.01172
Critical Values h,k   =          2.34, 1.75
```

71

P$_2$O$_5$, Cements E, Powder, Rep. 2

Laboratory Number	Cell Mean	Cell SD	*d*	*h*	*k*
1	0.1378	0.0010	0.0017	0.14	0.31
2	0.1400	0.0000	0.0039	0.30	0.00
3	0.1333	0.0052	-0.0028	-0.21	1.61
4	0.1433	0.0052	0.0072	0.56	1.61
5	0.1383	0.0041	0.0023	0.17	1.28
6	0.1353	0.0037	-0.0008	-0.06	1.16
7	0.1400	0.0000	0.0039	0.30	0.00
8	0.1000	0.0000	-0.0361	-2.80	0.00
9	0.1518	0.0012	0.0158	1.22	0.37
10	0.1377	0.0018	0.0016	0.12	0.55
11	0.1392	0.0048	0.0032	0.25	1.51

```
Average of cell averages           =        0.13608
Standard Deviation of cell averages =        0.01289
Repeatability Standard Deviation   =        0.00320
Reproducibility Standard Deviation =        0.01322
Critical Values h,k  =        2.34, 1.75
```

P$_2$O$_5$, Cements F, Powder, Rep. 2

Laboratory Number	Cell Mean	Cell SD	*d*	*h*	*k*
1	0.1365	0.0018	-0.0013	-0.11	0.51
2	0.1400	0.0000	0.0022	0.19	0.00
3	0.1333	0.0052	-0.0044	-0.38	1.50
4	0.1417	0.0041	0.0039	0.34	1.19
5	0.1400	0.0000	0.0022	0.19	0.00
6	0.1417	0.0055	0.0039	0.34	1.59
7	0.1417	0.0041	0.0039	0.34	1.19
8	0.1067	0.0052	-0.0311	-2.69	1.50
9	0.1543	0.0012	0.0166	1.43	0.35
10	0.1415	0.0015	0.0037	0.32	0.44
11	0.1381	0.0025	0.0004	0.03	0.72

```
Average of cell averages           =        0.13777
Standard Deviation of cell averages =        0.01154
Repeatability Standard Deviation   =        0.00343
Reproducibility Standard Deviation =        0.01196
Critical Values h,k  =        2.34, 1.75
```

XRF Glass, Replicate 1

Material	Xbar	s_x	s_r	s_R	r	R
1	0.1363	0.0038	0.0021	0.0042	0.01	0.01
2	0.1345	0.0035	0.0020	0.0040	0.01	0.01

XRF Glass, Replicate 2

Material	Xbar	s_x	s_r	s_R	r	R
1	0.1368	0.0041	0.0018	0.0044	0.00	0.01
2	0.1351	0.0030	0.0019	0.0035	0.01	0.01

XRF Powder Replicate 1

Material	Xbar	s_x	s_r	s_R	r	R
1	0.1368	0.0131	0.0028	0.0133	0.01	0.04
2	0.1375	0.0113	0.0033	0.0117	0.01	0.03

XRF Powder Replicate 2

Material	Xbar	s_x	s_r	s_R	r	R
1	0.1361	0.0129	0.0032	0.0132	0.01	0.04
2	0.1378	0.0115	0.0034	0.0120	0.01	0.03

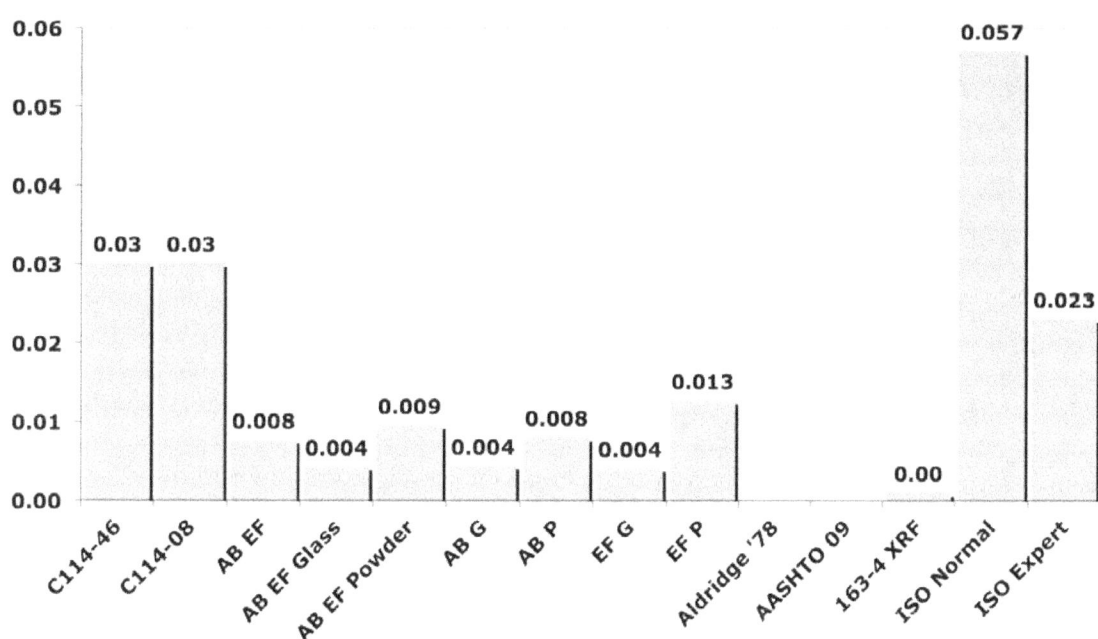

Figure 25 P_2O_5 precision statistics by method with bar chart comparing results to current and past ASTM C114 limits and previous studies on chemical analysis precision as 1σ, between lab (S_R).

Mn₂O₃

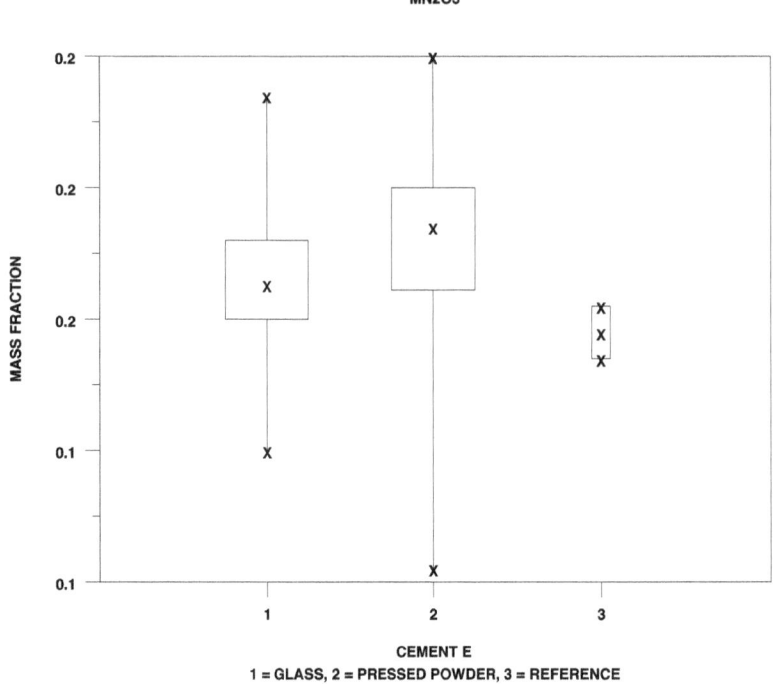

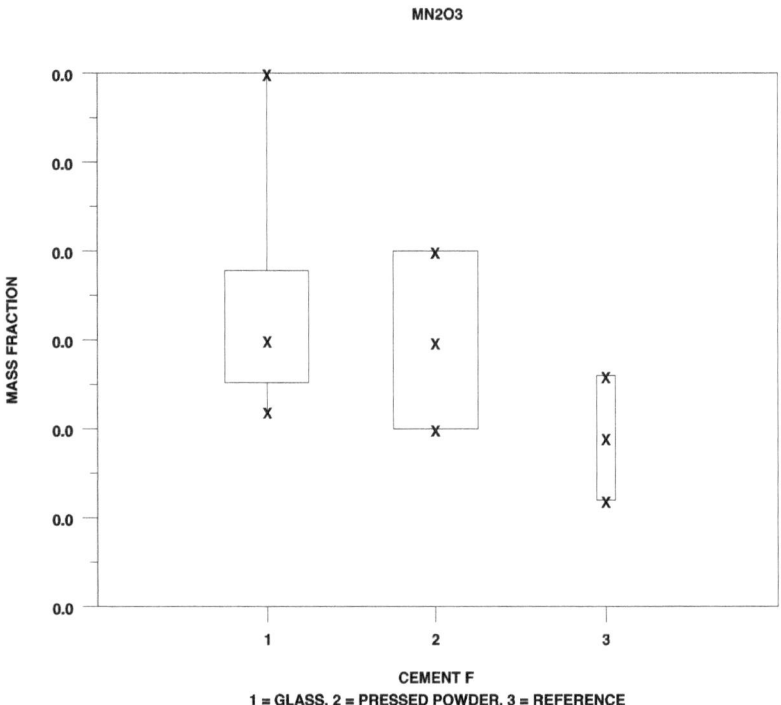

Figure 26 Box plots for Mn₂O₃ for XRF glass and powder, and reference methods.

Mn₂O₃, Cements E, Glass, Rep. 1

Laboratory Number	Cell Mean	Cell SD	d	h	k
1	0.2033	0.0029	-0.0006	-0.06	1.40
2	0.2100	0.0000	0.0060	0.61	0.00
3	0.2100	0.0000	0.0060	0.61	0.00
4	0.2073	0.0014	0.0034	0.34	0.65
5	0.2025	0.0009	-0.0015	-0.15	0.42
6	0.1830	0.0017	-0.0210	-2.12	0.80
7	0.2117	0.0041	0.0077	0.78	1.95

```
Average of cell averages           =        0.20397
Standard Deviation of cell averages  =        0.00988
Repeatability Standard Deviation   =        0.00210
Reproducibility Standard Deviation =        0.01006
Critical Values h,k   =          2.05, 1.70
```

Mn₂O₃, Cements F, Glass, Rep. 1

Laboratory Number	Cell Mean	Cell SD	d	h	k
1	0.0558	0.0023	0.0008	0.24	0.77
2	0.0600	0.0000	0.0050	1.51	0.00
3	0.0533	0.0052	-0.0017	-0.51	1.71
4	0.0582	0.0015	0.0031	0.95	0.49
5	0.0544	0.0016	-0.0007	-0.20	0.55
6	0.0502	0.0004	-0.0049	-1.48	0.14
7	0.0533	0.0052	-0.0017	-0.51	1.71

```
Average of cell averages           =        0.05503
Standard Deviation of cell averages  =        0.00329
Repeatability Standard Deviation   =        0.00302
Reproducibility Standard Deviation =        0.00429
Critical Values h,k   =          2.05, 1.70
```

Mn$_2$O$_3$, Cements E, Glass, Rep. 2

Laboratory Number	Cell Mean	Cell SD	d	h	k
1	0.2057	0.0016	0.0046	0.84	0.92
2	0.2067	0.0016	0.0056	1.02	0.92
3	0.2024	0.0011	0.0014	0.25	0.62
4	0.2022	0.0012	0.0011	0.20	0.66
5	0.2013	0.0029	0.0003	0.05	1.62
6	0.1990	0.0024	-0.0020	-0.36	1.37
7	0.1900	0.0000	-0.0110	-2.00	0.00

```
Average of cell averages          =        0.20104
Standard Deviation of cell averages  =     0.00551
Repeatability Standard Deviation  =        0.00178
Reproducibility Standard Deviation  =      0.00574
Critical Values h,k  =        2.05, 1.70
```

Mn$_2$O$_3$, Cements F, Glass, Rep. 2

Laboratory Number	Cell Mean	Cell SD	d	h	k
1	0.0535	0.0005	-0.0035	-0.58	0.39
2	0.0558	0.0023	-0.0012	-0.19	1.66
3	0.0537	0.0004	-0.0033	-0.55	0.28
4	0.0552	0.0004	-0.0018	-0.30	0.29
5	0.0527	0.0012	-0.0043	-0.72	0.87
6	0.0581	0.0025	0.0011	0.19	1.78
7	0.0700	0.0000	0.0130	2.16	0.00

```
Average of cell averages          =        0.05700
Standard Deviation of cell averages  =     0.00601
Repeatability Standard Deviation  =        0.00139
Reproducibility Standard Deviation  =      0.00615
Critical Values h,k  =        2.05, 1.70
```

Mn$_2$O$_3$, Cements E, Powder, Rep. 1

Laboratory Number	Cell Mean	Cell SD	*d*	*h*	*k*
1	0.2043	0.0015	0.0005	0.06	0.59
2	0.2100	0.0000	0.0062	0.72	0.00
3	0.2100	0.0000	0.0062	0.72	0.00
4	0.2082	0.0019	0.0043	0.50	0.76
5	0.2023	0.0033	-0.0016	-0.18	1.30
6	0.1855	0.0014	-0.0184	-2.14	0.54
7	0.2067	0.0052	0.0028	0.33	2.02

```
Average of cell averages          =        0.20385
Standard Deviation of cell averages  =      0.00858
Repeatability Standard Deviation     =      0.00255
Reproducibility Standard Deviation   =      0.00889
Critical Values h,k  =        2.05, 1.70
```

Mn$_2$O$_3$, Cements F, Powder, Rep. 1

Laboratory Number	Cell Mean	Cell SD	*d*	*h*	*k*
1	0.0555	0.0005	0.0012	0.33	0.22
2	0.0600	0.0000	0.0057	1.60	0.00
3	0.0517	0.0041	-0.0027	-0.75	1.67
4	0.0577	0.0018	0.0033	0.94	0.72
5	0.0535	0.0019	-0.0008	-0.24	0.78
6	0.0503	0.0012	-0.0040	-1.13	0.50
7	0.0517	0.0041	-0.0027	-0.75	1.67

```
Average of cell averages          =        0.05433
Standard Deviation of cell averages  =      0.00355
Repeatability Standard Deviation     =      0.00244
Reproducibility Standard Deviation   =      0.00419
Critical Values h,k  =        2.05, 1.70
```

Mn$_2$O$_3$, Cements E, Powder, Rep. 2

Laboratory Number	Cell Mean	Cell SD	d	h	k
1	0.2043	0.0015	0.0005	0.06	0.59
2	0.2100	0.0000	0.0062	0.72	0.00
3	0.2100	0.0000	0.0062	0.72	0.00
4	0.2082	0.0019	0.0043	0.50	0.76
5	0.2023	0.0033	-0.0016	-0.18	1.30
6	0.1855	0.0014	-0.0184	-2.14	0.54
7	0.2067	0.0052	0.0028	0.33	2.02

```
Average of cell averages          =        0.20385
Standard Deviation of cell averages  =     0.00858
Repeatability Standard Deviation  =        0.00255
Reproducibility Standard Deviation =       0.00889
Critical Values h,k  =        2.05, 1.70
```

Mn$_2$O$_3$, Cements F, Powder, Rep. 2

Laboratory Number	Cell Mean	Cell SD	d	h	k
1	0.0555	0.0005	0.0012	0.33	0.22
2	0.0600	0.0000	0.0057	1.60	0.00
3	0.0517	0.0041	-0.0027	-0.75	1.67
4	0.0577	0.0018	0.0033	0.94	0.72
5	0.0535	0.0019	-0.0008	-0.24	0.78
6	0.0503	0.0012	-0.0040	-1.13	0.50
7	0.0517	0.0041	-0.0027	-0.75	1.67

```
Average of cell averages          =        0.05433
Standard Deviation of cell averages  =     0.00355
Repeatability Standard Deviation  =        0.00244
Reproducibility Standard Deviation =       0.00419
Critical Values h,k  =        2.05, 1.70
```

XRF Glass, Replicate 1

Material	Xbar	s_x	s_r	s_R	r	R
1	0.2040	0.0099	0.0021	0.0101	0.01	0.03
2	0.0550	0.0033	0.0030	0.0043	0.01	0.01

XRF Glass, Replicate 2

Material	Xbar	s_x	s_r	s_R	r	R
1	0.2010	0.0055	0.0018	0.0057	0.00	0.02
2	0.0570	0.0060	0.0014	0.0061	0.00	0.02

XRF Powder Replicate 1

Material	Xbar	s_x	s_r	s_R	r	R
1	0.2039	0.0086	0.0026	0.0089	0.01	0.02
2	0.0543	0.0035	0.0024	0.0042	0.01	0.01

XRF Powder Replicate 2

Material	Xbar	s_x	s_r	s_R	r	R
1	0.2039	0.0086	0.0026	0.0089	0.01	0.02
2	0.0543	0.0035	0.0024	0.0042	0.01	0.01

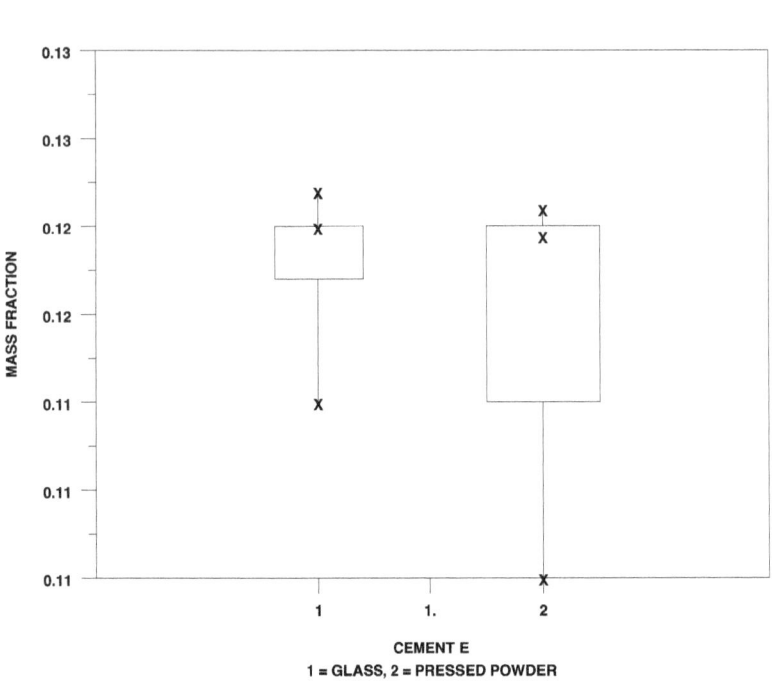

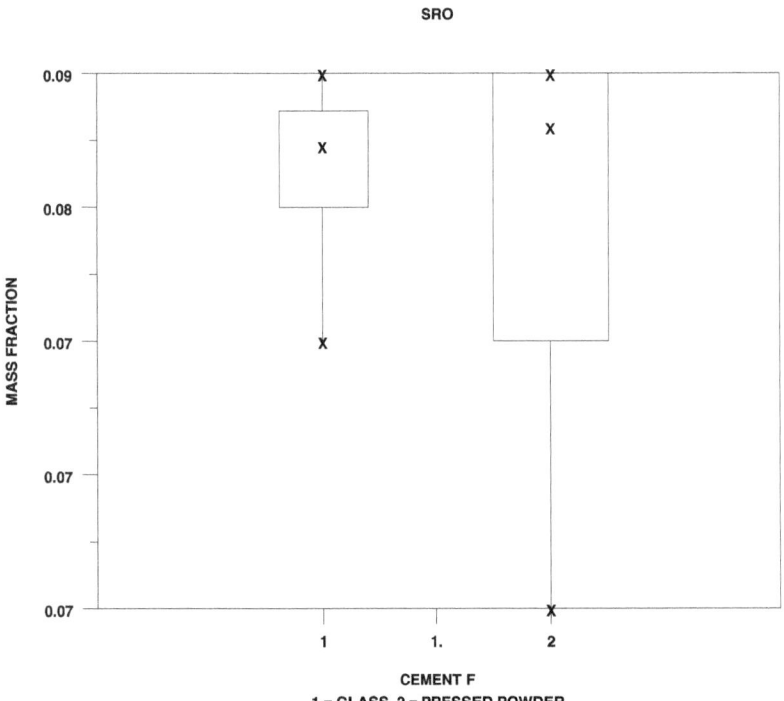

Figure 27 Box plots for SrO for XRF glass and powder methods.

SrO, Cements E, Glass, Rep. 1

Laboratory Number	Cell Mean	Cell SD	d	h	k
1	0.1272	0.0008	-0.0004	-0.10	1.12
2	0.1300	0.0000	0.0024	0.60	0.00
3	0.1306	0.0005	0.0030	0.74	0.77
4	0.1200	0.0000	-0.0076	-1.87	0.00
5	0.1307	0.0012	0.0031	0.76	1.81
6	0.1270	0.0006	-0.0005	-0.13	0.94

```
Average of cell averages           =        0.12757
Standard Deviation of cell averages =        0.00406
Repeatability Standard Deviation   =        0.00067
Reproducibility Standard Deviation =        0.00410
Critical Values h,k  =        1.92, 1.68
```

SrO, Cements F, Glass, Rep. 1

Laboratory Number	Cell Mean	Cell SD	d	h	k
1	0.0853	0.0005	-0.0007	-0.20	0.98
2	0.0900	0.0000	0.0040	1.13	0.00
3	0.0884	0.0004	0.0024	0.68	0.78
4	0.0800	0.0000	-0.0060	-1.72	0.00
5	0.0875	0.0005	0.0015	0.42	1.04
6	0.0849	0.0010	-0.0011	-0.31	1.83

```
Average of cell averages           =        0.08603
Standard Deviation of cell averages =        0.00351
Repeatability Standard Deviation   =        0.00053
Reproducibility Standard Deviation =        0.00355
Critical Values h,k  =        1.92, 1.68
```

SrO, Cements E, Glass, Rep. 2

Laboratory Number	Cell Mean	Cell SD	d	h	k
1	0.1275	0.0005	-0.0002	-0.05	0.67
2	0.1300	0.0000	0.0023	0.56	0.00
3	0.1307	0.0004	0.0030	0.74	0.50
4	0.1200	0.0000	-0.0077	-1.88	0.00
5	0.1308	0.0013	0.0031	0.76	1.63
6	0.1272	0.0013	-0.0005	-0.12	1.62

```
Average of cell averages           =         0.12771
Standard Deviation of cell averages =         0.00410
Repeatability Standard Deviation    =         0.00081
Reproducibility Standard Deviation  =         0.00416
Critical Values h,k  =        1.92, 1.68
```

SrO, Cements F, Glass, Rep. 2

Laboratory Number	Cell Mean	Cell SD	d	h	k
1	0.0855	0.0005	-0.0005	-0.14	1.18
2	0.0900	0.0000	0.0040	1.15	0.00
3	0.0884	0.0004	0.0024	0.70	0.88
4	0.0800	0.0000	-0.0060	-1.72	0.00
5	0.0872	0.0008	0.0012	0.34	1.62
6	0.0849	0.0005	-0.0011	-0.32	1.10

```
Average of cell averages           =         0.08599
Standard Deviation of cell averages =         0.00349
Repeatability Standard Deviation    =         0.00046
Reproducibility Standard Deviation  =         0.00351
Critical Values h,k  =        1.92, 1.68
```

SrO, Cements E, Powder, Rep. 1

Laboratory Number	Cell Mean	Cell SD	d	h	k
1	0.1300	0.0006	0.0029	0.64	0.39
2	0.1300	0.0000	0.0029	0.64	0.00
3	0.1300	0.0000	0.0029	0.64	0.00
4	0.1300	0.0000	0.0029	0.64	0.00
5	0.1200	0.0000	-0.0071	-1.59	0.00
6	0.1289	0.0005	0.0018	0.40	0.33
7	0.1198	0.0019	-0.0073	-1.62	1.19
8	0.1283	0.0041	0.0012	0.27	2.51

```
Average of cell averages            =        0.12714
Standard Deviation of cell averages =        0.00450
Repeatability Standard Deviation    =        0.00162
Reproducibility Standard Deviation  =        0.00474
Critical Values h,k  =        2.15, 1.72
```

SrO, Cements F, Powder, Rep. 1

Laboratory Number	Cell Mean	Cell SD	d	h	k
1	0.0877	0.0005	0.0033	0.49	0.25
2	0.0900	0.0000	0.0056	0.84	0.00
3	0.0850	0.0055	0.0006	0.09	2.60
4	0.0900	0.0000	0.0056	0.84	0.00
5	0.0700	0.0000	-0.0144	-2.15	0.00
6	0.0882	0.0002	0.0038	0.56	0.12
7	0.0843	0.0023	-0.0001	-0.01	1.07
8	0.0800	0.0000	-0.0044	-0.66	0.00

```
Average of cell averages            =        0.08440
Standard Deviation of cell averages =        0.00670
Repeatability Standard Deviation    =        0.00210
Reproducibility Standard Deviation  =        0.00697
Critical Values h,k  =        2.15, 1.72
```

SrO, Cements E, Powder, Rep. 2

Laboratory Number	Cell Mean	Cell SD	d	h	k
1	0.1302	0.0008	0.0031	0.73	0.40
2	0.1300	0.0000	0.0029	0.69	0.00
3	0.1300	0.0000	0.0029	0.69	0.00
4	0.1300	0.0000	0.0029	0.69	0.00
5	0.1200	0.0000	-0.0070	-1.65	0.00
6	0.1287	0.0005	0.0017	0.39	0.24
7	0.1208	0.0010	-0.0062	-1.46	0.52
8	0.1267	0.0052	-0.0004	-0.09	2.74

```
Average of cell averages            =        0.12705
Standard Deviation of cell averages =        0.00426
Repeatability Standard Deviation    =        0.00188
Reproducibility Standard Deviation  =        0.00460
Critical Values h,k  =        2.15, 1.72
```

SrO, Cements F, Powder, Rep. 2

Laboratory Number	Cell Mean	Cell SD	d	h	k
1	0.0882	0.0008	0.0031	0.45	0.48
2	0.0900	0.0000	0.0049	0.72	0.00
3	0.0883	0.0041	0.0033	0.47	2.63
4	0.0900	0.0000	0.0049	0.72	0.00
5	0.0700	0.0000	-0.0151	-2.18	0.00
6	0.0886	0.0004	0.0036	0.52	0.29
7	0.0853	0.0014	0.0003	0.04	0.88
8	0.0800	0.0000	-0.0051	-0.73	0.00

```
Average of cell averages            =        0.08506
Standard Deviation of cell averages =        0.00691
Repeatability Standard Deviation    =        0.00155
Reproducibility Standard Deviation  =        0.00705
Critical Values h,k  =        2.15, 1.72
```

XRF Glass, Replicate 1

Material	Xbar	s_x	s_r	s_R	r	R
1	0.1276	0.0041	0.0007	0.0041	0.00	0.01
2	0.0860	0.0035	0.0005	0.0035	0.00	0.01

XRF Glass, Replicate 2

Material	Xbar	s_x	s_r	s_R	r	R
1	0.1277	0.0041	0.0008	0.0042	0.00	0.01
2	0.0860	0.0035	0.0005	0.0035	0.00	0.01

XRF Powder Replicate 1

Material	Xbar	s_x	s_r	s_R	r	R
1	0.1271	0.0045	0.0016	0.0047	0.00	0.01
2	0.0844	0.0067	0.0021	0.0070	0.01	0.02

XRF Powder Replicate 2

Material	Xbar	s_x	s_r	s_R	r	R
1	0.1270	0.0043	0.0019	0.0046	0.01	0.01
2	0.0851	0.0069	0.0016	0.0071	0.00	0.02

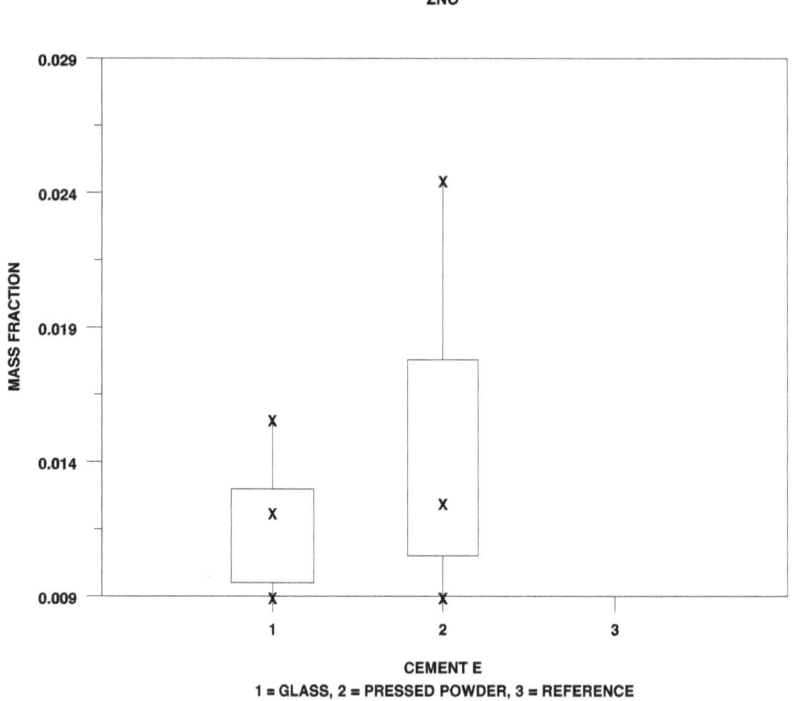

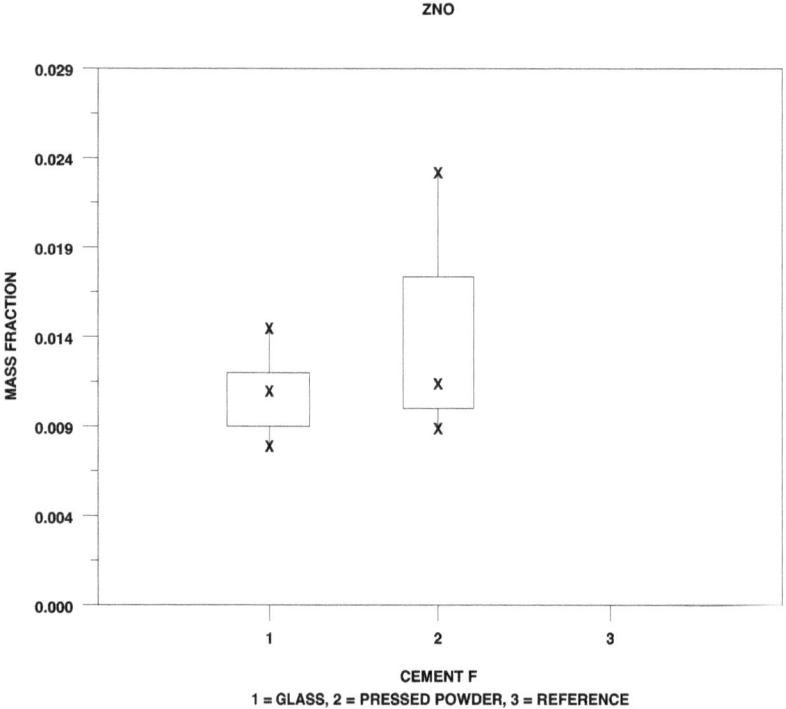

Figure 28 Box plots for ZnO for XRF glass and powder methods.

ZnO, Cement E, Glass and Powder, Rep. 1

Laboratory Number	Cell Mean	Cell SD	d	h	k
1	0.0140	0.0000	0.0001	0.02	0.00
2	0.0135	0.0005	-0.0004	-0.10	1.16
3	0.0164	0.0004	0.0025	0.58	0.86
4	0.0100	0.0000	-0.0039	-0.91	0.00
5	0.0130	0.0000	-0.0009	-0.21	0.00
6	0.0100	0.0000	-0.0039	-0.91	0.00
7	0.0247	0.0007	0.0108	2.51	1.45
8	0.0105	0.0005	-0.0034	-0.79	1.16
9	0.0129	0.0008	-0.0010	-0.23	1.62
10	0.0140	0.0006	0.0001	0.03	1.35

```
Average of cell averages            =        0.01391
Standard Deviation of cell averages =        0.00432
Repeatability Standard Deviation    =        0.00047
Reproducibility Standard Deviation  =        0.00434
Critical Values h,k  =         2.29, 1.74
```

ZnO, Cement F, Glass and Powder, Rep. 1

Laboratory Number	Cell Mean	Cell SD	d	h	k
1	0.0132	0.0004	0.0001	0.02	0.81
2	0.0127	0.0005	-0.0004	-0.10	1.03
3	0.0154	0.0004	0.0024	0.59	0.81
4	0.0100	0.0000	-0.0031	-0.76	0.00
5	0.0120	0.0000	-0.0011	-0.27	0.00
6	0.0100	0.0000	-0.0031	-0.76	0.00
7	0.0234	0.0006	0.0103	2.57	1.15
8	0.0097	0.0005	-0.0034	-0.85	1.03
9	0.0121	0.0010	-0.0010	-0.25	1.90
10	0.0124	0.0006	-0.0007	-0.18	1.27

```
Average of cell averages            =        0.01307
Standard Deviation of cell averages =        0.00402
Repeatability Standard Deviation    =        0.00050
Reproducibility Standard Deviation  =        0.00405
Critical Values h,k  =         2.29, 1.74
```

ZnO, Cement E, Glass and Powder, Rep. 2

Laboratory Number	Cell Mean	Cell SD	d	h	k
1	0.0140	0.0000	0.0001	0.02	0.00
2	0.0138	0.0004	-0.0001	-0.02	0.91
3	0.0164	0.0004	0.0025	0.58	0.91
4	0.0100	0.0000	-0.0039	-0.89	0.00
5	0.0132	0.0004	-0.0007	-0.17	0.91
6	0.0100	0.0000	-0.0039	-0.89	0.00
7	0.0250	0.0005	0.0111	2.53	1.14
8	0.0107	0.0005	-0.0032	-0.74	1.15
9	0.0127	0.0008	-0.0012	-0.28	1.69
10	0.0133	0.0006	-0.0006	-0.15	1.43

```
Average of cell averages            =         0.01390
Standard Deviation of cell averages =         0.00438
Repeatability Standard Deviation    =         0.00045
Reproducibility Standard Deviation  =         0.00440
Critical Values h,k  =        2.29, 1.74
```

ZnO, Cement F, Glass and Powder, Rep. 2

Laboratory Number	Cell Mean	Cell SD	d	h	k
1	0.0132	0.0004	0.0000	0.01	0.90
2	0.0128	0.0004	-0.0003	-0.07	0.90
3	0.0153	0.0005	0.0021	0.53	1.14
4	0.0100	0.0000	-0.0031	-0.77	0.00
5	0.0120	0.0000	-0.0011	-0.27	0.00
6	0.0100	0.0000	-0.0031	-0.77	0.00
7	0.0236	0.0007	0.0105	2.58	1.46
8	0.0097	0.0005	-0.0035	-0.85	1.14
9	0.0118	0.0005	-0.0013	-0.32	1.07
10	0.0128	0.0007	-0.0003	-0.07	1.57

```
Average of cell averages            =         0.01312
Standard Deviation of cell averages =         0.00406
Repeatability Standard Deviation    =         0.00045
Reproducibility Standard Deviation  =         0.00408
Critical Values h,k  =        2.29, 1.74
```

XRF Glass and Powder, Replicate 1

Material	Xbar	s_x	s_r	s_R	r	R
1	0.0139	0.0043	0.0005	0.0043	0.00	0.01
2	0.0131	0.0040	0.0005	0.0040	0.00	0.01

XRF Glass and Powder , Replicate 2

Material	Xbar	s_x	s_r	s_R	r	R
1	0.0139	0.0044	0.0004	0.0044	0.00	0.01
2	0.0131	0.0041	0.0005	0.0041	0.00	0.01

Cr_2O_3

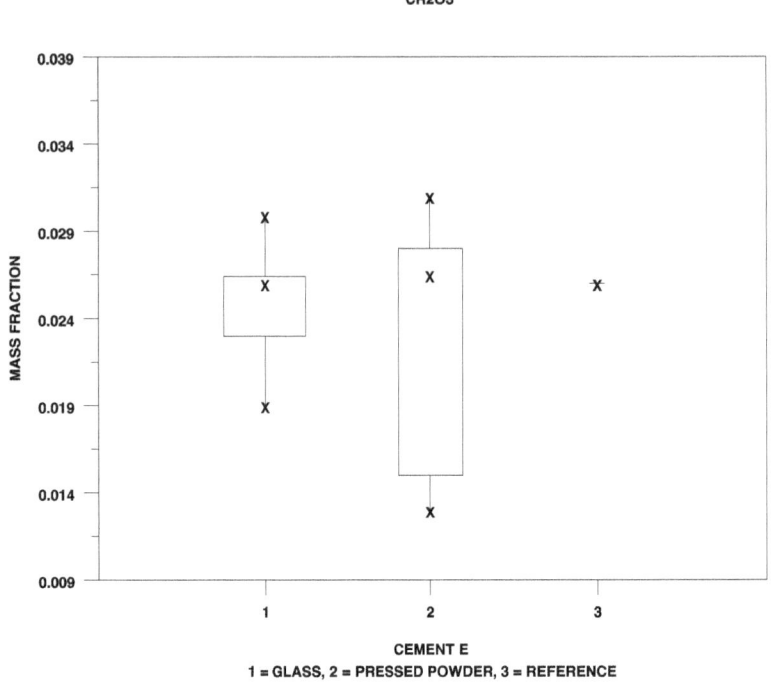

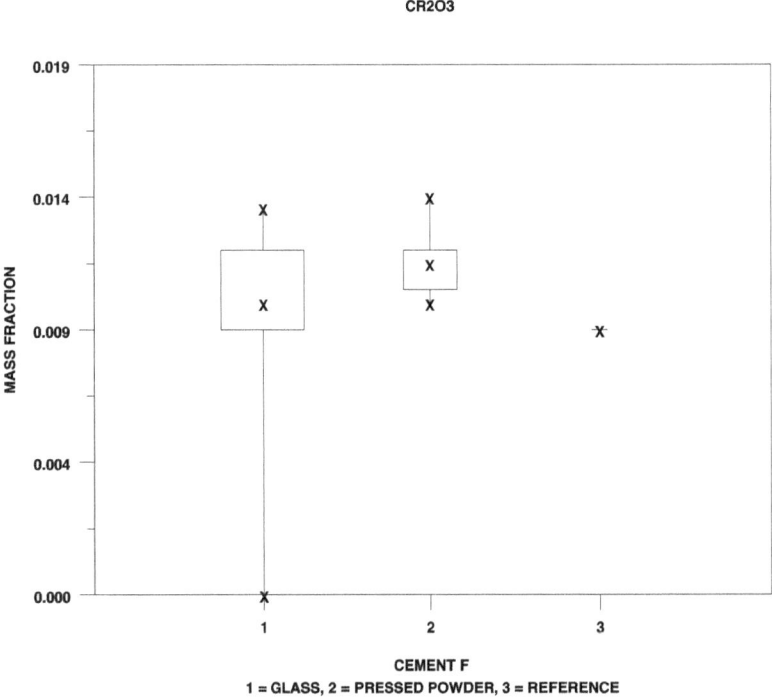

Figure 29 Box plots for Cr_2O_3 for XRF glass, powder, and reference methods.

Cr$_2$O$_3$, Cement E, Glass and Powder, Rep. 1

Laboratory Number	Cell Mean	Cell SD	d	h	k
1	0.0302	0.0012	0.0026	1.29	1.27
2	0.0272	0.0004	-0.0004	-0.21	0.44
3	0.0272	0.0004	-0.0004	-0.19	0.43
4	0.0275	0.0005	-0.0001	-0.04	0.60
5	0.0243	0.0008	-0.0033	-1.63	0.89
6	0.0291	0.0016	0.0016	0.78	1.69

```
Average of cell averages              =          0.02759
Standard Deviation of cell averages   =          0.00200
Repeatability Standard Deviation      =          0.00092
Reproducibility Standard Deviation    =          0.00217
Critical Values h,k   =        2.92, 1.68
```

Cr$_2$O$_3$, Cement F, Glass and Powder, Rep. 1

Laboratory Number	Cell Mean	Cell SD	d	h	k
1	0.0135	0.0008	0.0013	1.04	0.99
2	0.0112	0.0004	-0.0010	-0.79	0.49
3	0.0139	0.0005	0.0018	1.38	0.61
4	0.0115	0.0005	-0.0007	-0.53	0.65
5	0.0108	0.0013	-0.0013	-1.05	1.58
6	0.0121	0.0010	-0.0001	-0.05	1.22

```
Average of cell averages              =          0.01218
Standard Deviation of cell averages   =          0.00127
Repeatability Standard Deviation      =          0.00084
Reproducibility Standard Deviation    =          0.00149
Critical Values h,k   =        2.92, 1.68
```

Cr$_2$O$_3$, Cement E, Glass and Powder, Rep. 2

Laboratory Number	Cell Mean	Cell SD	d	h	k
1	0.0305	0.0005	0.0030	1.58	0.79
2	0.0270	0.0000	-0.0005	-0.27	0.00
3	0.0270	0.0005	-0.0005	-0.26	0.69
4	0.0277	0.0008	0.0002	0.08	1.18
5	0.0247	0.0005	-0.0028	-1.50	0.74
6	0.0282	0.0012	0.0007	0.36	1.72

```
Average of cell averages          =         0.02751
Standard Deviation of cell averages  =       0.00190
Repeatability Standard Deviation  =         0.00069
Reproducibility Standard Deviation   =       0.00200
Critical Values h,k  =        2.92, 1.68
```

Cr$_2$O$_3$, Cement F, Glass and Powder, Rep. 2

Laboratory Number	Cell Mean	Cell SD	d	h	k
1	0.0128	0.0004	0.0007	0.56	0.52
2	0.0110	0.0006	-0.0011	-0.86	0.80
3	0.0138	0.0004	0.0017	1.28	0.52
4	0.0115	0.0005	-0.0006	-0.47	0.69
5	0.0105	0.0012	-0.0016	-1.24	1.55
6	0.0131	0.0011	0.0010	0.74	1.40

```
Average of cell averages          =         0.01211
Standard Deviation of cell averages  =       0.00129
Repeatability Standard Deviation  =         0.00079
Reproducibility Standard Deviation   =       0.00148
Critical Values h,k  =        2.92, 1.68
```

XRF Glass and Powder Replicate 1

Material	Xbar	s_x	s_r	s_R	r	R
1	0.0276	0.0020	0.0009	0.0022	0.00	0.01
2	0.0122	0.0013	0.0008	0.0015	0.00	0.00

XRF Glass and Powder Replicate 2

Material	Xbar	s_x	s_r	s_R	r	R
1	0.0275	0.0019	0.0007	0.0020	0.00	0.01
2	0.0121	0.0013	0.0008	0.0015	0.00	0.00

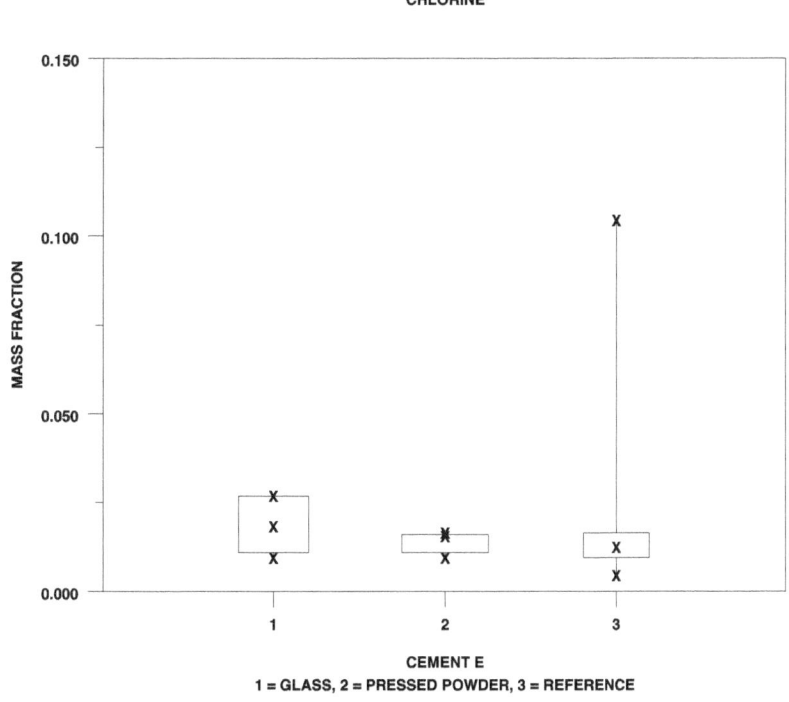

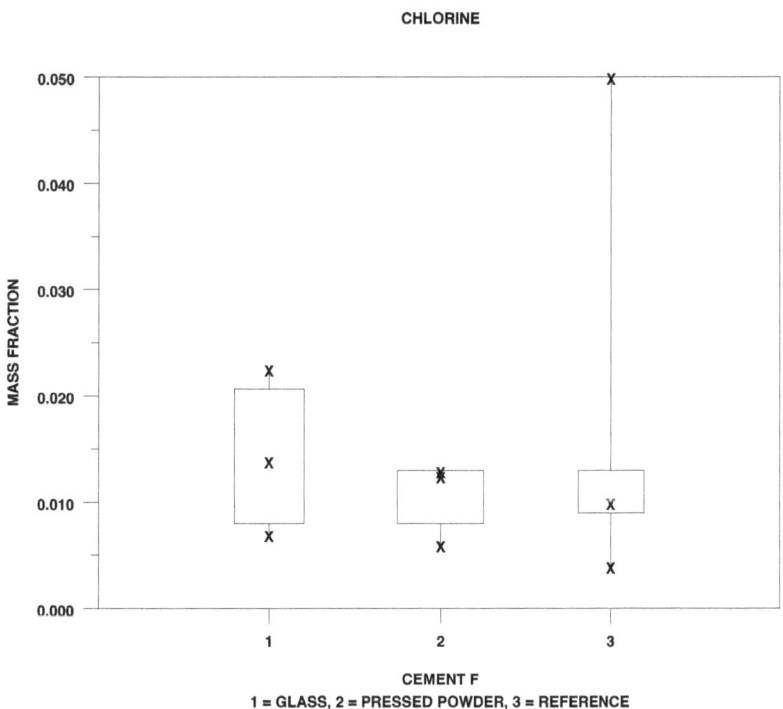

Figure 30 Box plots for Cl for XRF glass, powder, and reference methods.

Cl, Cement E, Glass and Powder, Rep. 1

Laboratory Number	Cell Mean	Cell SD	d	h	k
1	0.0158	0.0004	0.0008	0.13	0.73
2	0.0110	0.0006	-0.0040	-0.63	1.14
3	0.0100	0.0000	-0.0050	-0.79	0.00
4	0.0105	0.0005	-0.0045	-0.71	0.99
5	0.0160	0.0006	0.0010	0.15	1.14
6	0.0268	0.0008	0.0118	1.85	1.38

```
Average of cell averages           =        0.01502
Standard Deviation of cell averages =        0.00635
Repeatability Standard Deviation   =        0.00056
Reproducibility Standard Deviation =        0.00637
Critical Values h,k  =        1.92, 1.68
```

Cl, Cement F, Glass and Powder, Rep. 1

Laboratory Number	Cell Mean	Cell SD	d	h	k
1	0.0128	0.0004	0.0016	0.29	0.18
2	0.0077	0.0005	-0.0036	-0.67	0.23
3	0.0067	0.0052	-0.0046	-0.85	2.33
4	0.0072	0.0008	-0.0041	-0.76	0.34
5	0.0125	0.0008	0.0012	0.23	0.38
6	0.0208	0.0010	0.0095	1.76	0.47

```
Average of cell averages           =        0.01127
Standard Deviation of cell averages =        0.00540
Repeatability Standard Deviation   =        0.00222
Reproducibility Standard Deviation =        0.00577
Critical Values h,k  =        1.92, 1.68
```

Cl, Cement E, Glass and Powder, Rep. 2

Laboratory Number	Cell Mean	Cell SD	d	h	k
1	0.0162	0.0004	0.0009	0.13	0.45
2	0.0117	0.0008	-0.0036	-0.56	0.91
3	0.0100	0.0000	-0.0053	-0.81	0.00
4	0.0107	0.0008	-0.0046	-0.71	0.91
5	0.0158	0.0008	0.0005	0.08	0.84
6	0.0274	0.0017	0.0121	1.87	1.86

```
Average of cell averages              =          0.01530
Standard Deviation of cell averages   =          0.00650
Repeatability Standard Deviation      =          0.00090
Reproducibility Standard Deviation    =          0.00655
Critical Values h,k  =          1.92, 1.68
```

Cl, Cement F, Glass and Powder, Rep. 2

Laboratory Number	Cell Mean	Cell SD	d	h	k
1	0.0130	0.0000	0.0015	0.27	0.00
2	0.0082	0.0004	-0.0033	-0.60	0.19
3	0.0067	0.0052	-0.0048	-0.87	2.38
4	0.0073	0.0005	-0.0042	-0.75	0.24
5	0.0123	0.0005	0.0008	0.15	0.24
6	0.0215	0.0009	0.0100	1.80	0.41

```
Average of cell averages              =          0.01150
Standard Deviation of cell averages   =          0.00556
Repeatability Standard Deviation      =          0.00217
Reproducibility Standard Deviation    =          0.00591
Critical Values h,k  =          1.92, 1.68
```

XRF Glass and Powder Replicate 1

Material	Xbar	s_x	s_r	s_R	r	R
1	0.0150	0.0063	0.0006	0.0064	0.00	0.02
2	0.0113	0.0054	0.0022	0.0058	0.01	0.02

XRF Glass and Powder Replicate 2

Material	Xbar	s_x	s_r	s_R	r	R
1	0.0153	0.0065	0.0009	0.0066	0.00	0.02
2	0.0115	0.0056	0.0022	0.0059	0.01	0.02

Appendix A. Request for Participants Letter.

ASTM Round Robin for C01.23.03 Task Force Group

Instructions

The ASTM C01.23.03 task force group wishes to thank you for agreeing to be a part of the Round Robin (RR) which will define the precision and, hopefully, the bias of the X-ray fluorescence spectrometric (XRF) elemental analysis of hydraulic cements. Many versions of an XRF analytical procedure exist as set up and used by cement producers, cement users, commercial laboratories and general users. Generally, all parties are interested in producing a high quality, accurate analysis for industrial use. The Cement and Concrete Laboratory (CCRL) in cooperation with the ASTM Inter-Laboratory Study (ILS) has agreed to provide samples of Portland cement (meeting ASTM C150 guidelines) and various other mixtures of Portland cement and other compounds for the XRF analysis in this RR. Statistical analysis of the returned data will be compiled for the task force by NIST statisticians.

Method

The "10th Preliminary Draft" of "Standard Test Method for the Major and Minor Elements in Cement by X-Ray Fluorescence" produced by the task force group is available by down loading from a special web-based program set up by CCRL for the specific purpose of this RR.   The method is <u>specifically, a generalized</u> approach to the XRF analysis of hydraulic cements, and is subject to revision when data is collected to show any confusion, indecision, or conflicts in the "10th Preliminary Draft." There have already been "lots" of "discussions" and changes to the method, and there may be more changes. Some have already been suggested by task force members and other interested parties that may be implemented when actual data begins to roll in.

In the interest of collecting data from the RR before I go to a nursing home, we should start to collect now. Again the method is a generalized approach to XRF analysis of hydraulic cements, and the individual Standard Operating Procedures (SOP) used in your individual laboratory should fit in the generalized guidelines. Please follow the method as closely as possible so that your specific laboratory data can be utilized for comparison to all the other data.

The method actually has two preparation schemes for presentation to the XRF, <u>fused disk</u> and <u>pressed powder</u>. If your laboratory has the capability to do both preparation methods, please, please, do so.

In Part 4.2.2 of the "10th Preliminary Draft," it specifically states *"Part 2 of the method is not generally applicable to the analysis of blended cements, including those blended with silica fume, limestone, pozzolans, fly ash, slag, and any combination thereof higher than the amounts of additions found in Portland cements meeting Specification C 150."* **<u>For the purposes of this RR data collection only, ignore this statement and run pressed pellets, if you have the capability, for comparison with the fused disk preparation.</u>**   We need to establish the precision and truth of this statement and a comparison will determine if this statement is correct and help establish precision data on the pressed pellet mixtures by XRF.

Data Return

The data return for this RR is by accessing the CCRL web site for examination and determining what needs to be reported. It will be much easier if you report data on-line in a similar fashion as CCRL data is now reported; the statistician will love you. If you are unable to report data on-line, fill out the data and send to me (by FAX) and I'll key punch it in for your data return. By making a request to CCRL, on this RR you can obtain more sample, if you are going to do samples by pressed powder Ask for three vials of about 30 grams on each of the RR samples for powder prep and one sample vial if sample prep is fused disk only.

If you have the capability, we ask you to run powder prep and fused pellet on the two samples being mailed from CCRL. So for instance, there will be Sample A fused pellet data return and Sample A pressed pellet data return; then Sample B fused pellet data return and Sample B pressed pellet data return. There will be potentially 4 data return sheets for each sample in the mailing from CCRL. The first two samples mailed from CCRL are guaranteed to be C150 Portland Cements. The next two to four mailings of sample pairs will be mixtures.

If you can't do fused pellets, please participate in the pressed pellet RR, and conversely if you can't do pressed pellets, do fused pellets. All data will be used and gratefully appreciated by the task force group and task group chairman. We estimate that this RR study will take a minimum of 1½ years to 2½ years, a mere moment in the large ASTM scheme of things, but the data will hopefully live forever.

1. The section at the top should be filled out for company and individual returning the data, but, in general, no data will be identified with the lab unless you wish it to be. This ID is in case we have to contact you to clear up some misunderstanding or finger trouble with the data entry.

2. Please give as much data on the Fusion Bead preparation as possible. Flux composition, Non-wetting agent, etc. For the Pressed Pellet, as much info as possible, Binder composition, Grinding aid and etc.

3. Look at the return data sheet, what is intended here is to make on Day 1, three pellets: fusion and/or pressed pellets, and analyze each pellet twice, then report on the properly identified data sheets. Day 2 of your choosing, make three pellets: fusion and/or pressed pellets, and run each one twice, and report results on the proper data return sheets.

4. If possible (and only if possible) on your instrument, give the raw element intensities and/or raw background intensities before calculations and applied corrections. In the case of each element you only need to answer the Calibration/correction model question once, not every time on each prep.

5. LOI should be run and reported each sample prep day. Insoluble Residue and Free Lime need only to be run once, each prep day and only if the sufficient data is returned for the XRF data. The RR is primarily an XRF RR; the other data at the end of the CCRL data return will be used if needed.

6. Data should be reported to the CCRL web site with one more significant digit than the smallest uncertainty (or standard deviation) value of your best standard elemental value for the element being reported back to CCRL. Report determined values no later than three (3) months after receipt of samples in your laboratory. (i. e., SRM 1880a has a certified value for CaO of 63.83% +/- 0.46, report to three decimals for CaO unless you have lower values on reference materials that were used. ZnO is certified at 0.005% +/- 0.001, so report to four decimals in weight percent, unless you have used reference standards which have a lower +/- values.)

LeRoy Jacobs
ASTM C01.23.03 Task Force Group Chairman
Wyoming Analytical Lab, Inc.
1511 Washington Ave.
Golden, CO 80401
Ph (303) 278-2446
FAX (303) 278-2439
e:mail walxray@aol.com

Appendix B.   XRF Glass Summary by Material and Replicate.

Analyte	Material	Xbar	s_x	s_r	s_R	r	R	n
CaO	1	62.4493	0.4181	0.0843	0.4252	0.24	1.19	8
	2	63.2911	0.235	0.0879	0.2484	0.25	0.7	
	1	62.4455	0.3861	0.1162	0.4004	0.33	1.12	
	2	63.3026	0.261	0.0946	0.275	0.27	0.77	
SiO$_2$	1	20.5726	0.1682	0.0616	0.1773	0.17	0.5	9
	2	18.841	0.1457	0.0508	0.1529	0.14	0.43	
	1	20.5816	0.1796	0.0587	0.1874	0.16	0.52	
	2	18.857	0.1492	0.0502	0.1561	0.14	0.44	
Al$_2$O$_3$	1	4.445	0.0826	0.0286	0.0866	0.08	0.24	10
	2	5.3365	0.0624	0.0275	0.0672	0.08	0.19	
	1	4.4443	0.0765	0.0244	0.0797	0.07	0.22	
	2	5.3285	0.0614	0.0254	0.0656	0.07	0.18	
Fe$_2$O$_3$	1	2.8796	0.0409	0.0078	0.0415	0.02	0.12	10
	2	2.3715	0.0384	0.0148	0.0407	0.04	0.11	
	1	2.8854	0.0368	0.0092	0.0378	0.03	0.11	
	2	2.3794	0.0338	0.0092	0.0348	0.03	0.1	
SO$_3$	1	3.2355	0.0544	0.0129	0.0557	0.04	0.16	9
	2	3.6822	0.0802	0.0323	0.0855	0.09	0.24	
	1	3.2339	0.0533	0.0138	0.0548	0.04	0.15	
	2	3.6691	0.0822	0.0857	0.1135	0.24	0.32	
MgO	1	2.5645	0.0172	0.0132	0.021	0.04	0.06	9
	2	2.1092	0.0161	0.0095	0.0183	0.03	0.05	
	1	2.5647	0.0184	0.0114	0.0211	0.03	0.06	
	2	2.1105	0.0143	0.0101	0.017	0.03	0.05	
Na$_2$O	1	0.1545	0.0163	0.007	0.0175	0.02	0.05	10
	2	0.1447	0.0145	0.0059	0.0155	0.02	0.04	
	1	0.1552	0.0148	0.005	0.0155	0.01	0.04	
	2	0.1467	0.0135	0.0059	0.0145	0.02	0.04	
K$_2$O	1	0.7325	0.0082	0.0029	0.0086	0.01	0.02	11
	2	1.1742	0.0156	0.0049	0.0162	0.01	0.05	
	1	0.731	0.0062	0.0037	0.007	0.01	0.02	
	2	1.1753	0.0162	0.0047	0.0168	0.01	0.05	
TiO$_2$	1	0.2315	0.0078	0.0039	0.0085	0.01	0.02	12
	2	0.2262	0.0083	0.0032	0.0088	0.01	0.02	
	1	0.2314	0.0074	0.004	0.0083	0.01	0.02	
	2	0.2265	0.0075	0.0034	0.0082	0.01	0.02	

Analyte (continued)	Material	Xbar	s_x	s_r	s_R	r	R	n
P_2O_5	1	0.1363	0.0038	0.0021	0.0042	0.01	0.01	9
	2	0.1345	0.0035	0.002	0.004	0.01	0.01	
	1	0.1368	0.0041	0.0018	0.0044	0	0.01	
	2	0.1351	0.003	0.0019	0.0035	0.01	0.01	
Mn_2O_3	1	0.204	0.0099	0.0021	0.0101	0.01	0.03	7
	2	0.055	0.0033	0.003	0.0043	0.01	0.01	
	1	0.201	0.0055	0.0018	0.0057	0	0.02	
	2	0.057	0.006	0.0014	0.0061	0	0.02	
SrO	1	0.1276	0.0041	0.0007	0.0041	0	0.01	6
	2	0.086	0.0035	0.0005	0.0035	0	0.01	
	1	0.1277	0.0041	0.0008	0.0042	0	0.01	
	2	0.086	0.0035	0.0005	0.0035	0	0.01	
ZnO glass and powder	1	0.0139	0.0043	0.0005	0.0043	0	0.01	10
	2	0.0131	0.004	0.0005	0.004	0	0.01	
Cr_2O_3 glass and powder	1	0.0276	0.002	0.0009	0.0022	0	0.01	6
	2	0.0122	0.0013	0.0008	0.0015	0	0	
Cl glass and powder								6

Appendix C. XRF Powder Summary by Material and Replicate.

Analyte	Material	Xbar	s_x	s_r	s_R	r	R	n
CaO	1	62.6901	0.1602	0.1058	0.1871	0.3	0.52	9
	2	63.7063	0.3184	0.138	0.3424	0.39	0.96	
	1	62.7128	0.1569	0.1617	0.2154	0.45	0.6	
	2	63.7236	0.3305	0.133	0.3521	0.37	0.99	
SiO$_2$	1	20.5576	0.1928	0.0621	0.201	0.17	0.56	11
	2	19.1546	0.2818	0.0497	0.2854	0.14	0.8	
	1	20.575	0.1878	0.062	0.1961	0.17	0.55	
	2	19.1653	0.2976	0.0578	0.3022	0.16	0.85	
Al$_2$O$_3$	1	4.5251	0.1019	0.0235	0.1042	0.07	0.29	10
	2	5.1185	0.0968	0.0252	0.0995	0.07	0.28	
	1	4.5154	0.0882	0.0234	0.0907	0.07	0.25	
	2	5.1275	0.0971	0.0247	0.0997	0.07	0.28	
Fe$_2$O$_3$	1	2.9215	0.0409	0.014	0.0428	0.04	0.12	10
	2	2.3874	0.0446	0.0127	0.0461	0.04	0.13	
	1	2.9208	0.0363	0.0133	0.0383	0.04	0.11	
	2	2.3885	0.0465	0.0133	0.0481	0.04	0.13	
SO$_3$	1	3.1821	0.0872	0.0233	0.0898	0.07	0.25	11
	2	3.7005	0.1161	0.0325	0.1198	0.09	0.34	
	1	3.1702	0.1027	0.0231	0.1049	0.06	0.29	
	2	3.7002	0.1177	0.0336	0.1216	0.09	0.34	
MgO	1	2.6521	0.0528	0.0259	0.0578	0.07	0.16	12
	2	2.0619	0.0297	0.0152	0.0327	0.04	0.09	
	1	2.6495	0.0509	0.0238	0.0554	0.07	0.15	
	2	2.0652	0.0263	0.0162	0.0302	0.05	0.08	
Na$_2$O	1	0.159	0.0201	0.0071	0.0211	0.02	0.06	10
	2	0.1579	0.0219	0.0059	0.0225	0.02	0.06	
	1	0.1598	0.0197	0.0077	0.021	0.02	0.06	
	2	0.1571	0.0209	0.007	0.0218	0.02	0.06	
K$_2$O	1	0.7372	0.0083	0.0053	0.0096	0.01	0.03	10
	2	1.1896	0.0217	0.0078	0.0228	0.02	0.06	
	1	0.7369	0.008	0.005	0.0092	0.01	0.03	
	2	1.1895	0.0218	0.0096	0.0235	0.03	0.07	
TiO$_2$	1	0.2301	0.004	0.0033	0.005	0.01	0.01	12
	2	0.2215	0.006	0.0043	0.0072	0.01	0.02	
	1	0.23	0.0044	0.0031	0.0052	0.01	0.01	
	2	0.2213	0.006	0.0039	0.007	0.01	0.02	

Analyte (continued)	Material	Xbar	s_x	s_r	s_R	r	R	n
P_2O_5	1	0.1368	0.0131	0.0028	0.0133	0.01	0.04	11
	2	0.1375	0.0113	0.0033	0.0117	0.01	0.03	
	1	0.1361	0.0129	0.0032	0.0132	0.01	0.04	
	2	0.1378	0.0115	0.0034	0.012	0.01	0.03	
Mn_2O_3	1	0.2039	0.0086	0.0026	0.0089	0.01	0.02	7
	2	0.0543	0.0035	0.0024	0.0042	0.01	0.01	
	1							
	2							
SrO	1	0.1271	0.0045	0.0016	0.0047	0	0.01	8
	2	0.0844	0.0067	0.0021	0.007	0.01	0.02	
	1	0.127	0.0043	0.0019	0.0046	0.01	0.01	
	2	0.0851	0.0069	0.0016	0.0071	0	0.02	

Appendix D. Raw Data by Analyte.

Data by oxide after initial outlier removal for SiO_2, Al_2O_3, Fe_2O_3, CaO, MgO, SO_3, Na_2O, K_2O, TiO_2, and Cl. Data are organized with the XRF-glass results to the left and XRF-powder to the right. A lab identifier is provided in the first column of each preparation method and the six replicate measurements for that lab and analyte for duplicate 1 and then duplicate two. Duplicate measurements represent a repeat measurement on the same specimen. The final column labeled cement indicates cement E (1) or cement F (2). Since outliers were evaluated on an analyte-by analyte basis, a specific lab designation may not be the same lab across all analytes. For example, if lab 1 had a high between-lab precision for Al_2O_3, its data were removed and the labs reordered so no gap in the numbering existed.

SiO_2

Lab	SiO₂: Glass Preparation Duplicate 1	Duplicate 2	Cement	Lab	SiO₂: Powder Preparation Duplicate 1	Duplicate 2	Cement
1	20.225	20.295	1	1	20.74	20.73	1
1	20.458	20.489	1	1	20.6	20.72	1
1	20.489	20.514	1	1	20.58	20.63	1
1	20.447	20.469	1	1	20.69	20.69	1
1	20.495	20.54	1	1	20.66	20.72	1
1	20.59	20.653	1	1	20.61	20.67	1
2	20.825	20.605	1	2	20.88	20.87	1
2	20.568	20.337	1	2	20.83	20.88	1
2	20.728	20.615	1	2	20.77	20.78	1
2	20.337	20.475	1	2	20.91	20.89	1
2	20.244	20.22	1	2	20.82	20.75	1
2	20.558	20.681	1	2	20.81	20.76	1
3	20.81	20.91	1	3	21.08	21.14	1
3	20.74	20.85	1	3	21.12	21.14	1
3	20.83	20.81	1	3	21.45	21.43	1
3	20.82	20.84	1	3	20.56	20.64	1
3	20.76	20.83	1	3	19.78	19.81	1
3	20.86	20.99	1	3	20.26	20.38	1
4	20.328	20.317	1	4	20.73	20.81	1
4	20.32	20.292	1	4	20.7	20.69	1
4	20.302	20.2654	1	4	20.71	20.74	1
4	20.213	20.218	1	4	20.76	20.79	1
4	20.253	20.273	1	4	20.75	20.78	1
4	20.238	20.24	1	4	20.74	20.75	1
5	20.457	20.505	1	5	20.33	20.29	1
5	20.476	20.523	1	5	20.37	20.39	1
5	20.507	20.464	1	5	20.43	20.38	1
5	20.519	20.544	1	5	20.45	20.52	1
5	20.525	20.51	1	5	20.47	20.49	1
5	20.514	20.484	1	5	20.41	20.44	1
6	20.69	20.69	1	6	20.64	20.62	1
6	20.63	2064	1	6	20.75	20.75	1
6	20.72	20.71	1	6	20.61	20.66	1
6	20.71	20.71	1	6	20.54	20.57	1
6	20.68	20.68	1	6	20.62	20.64	1

6	20.72	20.73	1	6	20.57	20.55	1
7	20.71	20.64	1	7	20.78	20.86	1
7	20.69	20.69	1	7	20.88	20.89	1
7	20.64	20.65	1	7	20.99	20.98	1
7	20.6	20.67	1	7	20.69	20.69	1
7	20.72	20.62	1	7	20.89	20.88	1
7	20.77	20.74	1	7	20.82	20.83	1
8	20.47	20.52	1	8	20.48	20.5	1
8	20.55	20.54	1	8	20.52	20.56	1
8	20.56	20.54	1	8	20.47	20.49	1
8	20.55	20.48	1	8	20.56	20.53	1
8	20.52	20.52	1	8	20.56	20.57	1
8	20.53	20.5	1	8	20.57	20.56	1
9	20.745	20.7473	1	9	20.36	20.38	1
9	20.699	20.711	1	9	20.37	20.4	1
9	20.609	20.628	1	9	20.3	20.3	1
9	20.819	20.833	1	9	20.24	20.22	1
9	20.854	20.847	1	9	20.34	20.36	1
9	20.686	20.693	1	9	20.4	20.39	1
10	21.28	20.62	1	10	20.48	20.5	1
10	20.59	20.81	1	10	20.49	20.5	1
10	20.77	20.62	1	10	20.45	20.49	1
10	21.32	20.64	1	10	20.43	20.44	1
10	20.68	20.84	1	10	20.53	20.54	1
10	20.75	20.59	1	10	20.52	20.56	1
11	20.479	20.462	1	11	20.8	20.79	1
11	20.495	20.489	1	11	20.71	20.72	1
11	20.453	20.446	1	11	20.76	20.77	1
11	20.472	20.491	1	11	20.78	20.76	1
11	20.492	20.49	1	11	20.68	20.69	1
11	20.482	20.475	1	11	20.73	20.73	1
12	21.47	21.43	1	12	20.4	20.4	1
12	21.45	21.47	1	12	20.43	20.46	1
12	21.46	21.43	1	12	20.47	20.47	1
12	21.38	21.43	1	12	20.34	20.31	1
12	21.39	21.4	1	12	20.39	20.42	1
12	21.42	21.43	1	12	20.42	20.4	1
1	18.724	18.773	2	13	20.856	20.868	1
1	18.745	18.815	2	13	20.883	20.934	1
1	18.783	18.814	2	13	21.053	21.065	1
1	18.607	18.666	2	13	20.795	20.819	1
1	18.812	18.802	2	13	20.852	20.859	1
1	18.902	18.956	2	13	20.775	20.803	1
2	18.759	18.661	2	14	20.195	20.39	1
2	18.831	18.765	2	14	20.36	20.375	1
2	18.822	18.776	2	14	20.335	20.3	1
2	19.13	18.975	2	14	20.165	20.395	1
2	19.003	18.96	2	14	20.39	20.385	1
2	18.802	18.807	2	14	20.305	20.275	1
3	19	18.98	2	1	19.05	19.06	2

3	18.89	18.93	2	1	19.07	19.06	2
3	18.98	18.96	2	1	19.07	19.05	2
3	18.92	18.96	2	1	19.03	19.14	2
3	18.97	18.97	2	1	19.03	19.01	2
3	18.89	19.05	2	1	19.04	19.05	2
4	18.583	18.635	2	2	19.25	19.24	2
4	18.594	18.597	2	2	19.2	19.16	2
4	18.531	18.548	2	2	19.21	19.21	2
4	18.549	18.532	2	2	19.13	19.15	2
4	18.564	18.526	2	2	19.13	19.13	2
4	18.533	18.55	2	2	19.03	18.97	2
5	18.787	18.807	2	3	19.23	19.26	2
5	18.799	18.836	2	3	19.1	18.96	2
5	18.824	18.888	2	3	19.18	19.29	2
5	18.878	18.863	2	3	19	19.01	2
5	18.825	18.897	2	3	19.3	19.38	2
5	18.838	18.797	2	3	19.23	19.3	2
6	19.05	19.03	2	4	19.16	19.14	2
6	18.99	18.99	2	4	19.12	19.14	2
6	19.09	19.05	2	4	19.03	19.05	2
6	19	19.02	2	4	19.14	19.15	2
6	19	19	2	4	19.04	19.06	2
6	19.01	19.04	2	4	19.05	19.07	2
7	18.89	18.89	2	5	18.82	18.77	2
7	18.92	18.93	2	5	18.87	18.83	2
7	18.87	18.99	2	5	18.82	18.84	2
7	18.92	19.02	2	5	18.93	18.91	2
7	18.97	18.97	2	5	18.89	18.8	2
7	18.96	19.01	2	5	18.88	18.89	2
8	18.78	18.77	2	6	19.15	19.16	2
8	18.79	18.76	2	6	19.23	19.35	2
8	18.78	18.83	2	6	19.17	19.17	2
8	18.75	18.76	2	6	19.13	19.15	2
8	18.79	18.79	2	6	19.06	19.18	2
8	18.81	18.83	2	6	19.11	19.12	2
9	18.985	18.967	2	7	19.38	19.38	2
9	18.942	18.953	2	7	19.29	19.33	2
9	18.904	18.927	2	7	19.29	19.28	2
9	19.085	19.074	2	7	19.32	19.34	2
9	18.998	18.994	2	7	19.34	19.36	2
9	19.055	19.062	2	7	19.19	19.21	2
10	18.94	18.97	2	8	19.11	19.08	2
10	19.01	19.03	2	8	19.02	19.02	2
10	18.92	18.86	2	8	19	19.05	2
10	18.92	18.97	2	8	19.08	19.05	2
10	19.01	19	2	8	19.05	19.06	2
10	18.89	18.89	2	8	19.06	19.03	2
11	18.753	18.758	2	9	19.18	19.16	2
11	18.79	18.791	2	9	19.11	19.14	2
11	18.804	18.771	2	9	19.16	19.16	2

11	18.767	18.714	2	9	19.14	19.16	2
11	18.689	18.706	2	9	19.05	19.06	2
11	18.746	18.73	2	9	19.16	19.16	2
12	19.76	19.66	2	10	18.72	18.75	2
12	19.76	19.73	2	10	18.72	18.74	2
12	19.7	19.75	2	10	18.75	18.77	2
12	19.67	19.7	2	10	18.8	18.82	2
12	19.68	19.61	2	10	18.75	18.78	2
12	19.65	19.71	2	10	18.74	18.77	2
				11	19.6	19.61	2
				11	19.51	16.61	2
				11	19.56	19.58	2
				11	19.58	19.58	2
				11	19.49	19.5	2
				11	19.55	19.55	2
				12	19.34	19.3	2
				12	19.34	19.32	2
				12	19.38	19.34	2
				12	19.3	19.29	2
				12	19.26	19.27	2
				12	19.29	19.26	2
				13	19.704	19.713	2
				13	19.58	19.585	2
				13	19.569	19.564	2
				13	19.367	19.38	2
				13	19.406	19.419	2
				13	19.515	19.519	2
				14	19.885	19.895	2
				14	19.85	20.015	2
				14	19.785	19.83	2
				14	19.865	19.91	2
				14	19.865	19.995	2
				14	19.795	19.815	2

Al2O3: Glass Preparation				Al2O3: Powder Preparation			
Lab	Duplicate 1	Duplicate 2	Cement	Lab	Duplicate 1	Duplicate 2	Cement
1	4.355	4.366	1	1	4.49	4.45	1
1	4.402	4.411	1	1	4.46	4.49	1
1	4.403	4.403	1	1	4.48	4.47	1
1	4.386	4.393	1	1	4.46	4.44	1
1	4.418	4.428	1	1	4.47	4.43	1
1	4.431	4.435	1	1	4.46	4.41	1
2	4.678	4.654	1	2	4.53	4.53	1
2	4.655	4.562	1	2	4.55	4.55	1
2	4.624	4.598	1	2	4.55	4.56	1
2	4.533	4.547	1	2	4.56	4.56	1
2	4.578	4.575	1	2	4.53	4.53	1
2	4.588	4.59	1	2	4.54	4.53	1
3	4.51	4.49	1	3	4.93	4.95	1
3	4.45	4.51	1	3	4.98	4.94	1
3	4.49	4.49	1	3	4.85	4.87	1
3	4.51	4.49	1	3	4.76	4.79	1
3	4.49	4.47	1	3	4.72	4.74	1
3	4.53	4.51	1	3	4.78	4.8	1
4	4.4479	4.4557	1	4	4.52	4.5	1
4	4.4293	4.4243	1	4	4.53	4.51	1
4	4.4262	4.4341	1	4	4.52	4.51	1
4	4.4265	4.4148	1	4	4.52	4.51	1
4	4.4374	4.4326	1	4	4.51	4.53	1
4	4.4168	4.4159	1	4	4.5	4.51	1
5	4.512	4.505	1	5	4.54	4.54	1
5	4.448	4.483	1	5	4.55	4.55	1
5	4.462	4.494	1	5	4.61	4.62	1
5	4.498	4.464	1	5	4.51	4.52	1
5	4.463	4.477	1	5	4.49	4.49	1
5	4.504	4.481	1	5	4.48	4.47	1
6	4.42	4.43	1	6	4.49	4.5	1
6	4.46	4.47	1	6	4.48	4.48	1
6	4.45	4.45	1	6	4.52	4.5	1
6	4.43	4.41	1	6	4.53	4.53	1
6	4.46	4.44	1	6	4.5	4.51	1
6	4.45	4.45	1	6	4.52	4.5	1
7	4.34	4.31	1	7	4.57	4.54	1
7	4.27	4.25	1	7	4.55	4.55	1
7	4.33	4.3	1	7	4.53	4.54	1
7	4.29	4.33	1	7	4.52	4.52	1
7	4.27	4.25	1	7	4.55	4.51	1
7	4.31	4.34	1	7	4.55	4.56	1
8	4.468	4.4651	1	8	4.41	4.44	1
8	4.4942	4.5033	1	8	4.41	4.42	1
8	4.4654	4.459	1	8	4.41	4.44	1
8	4.4998	4.5137	1	8	4.29	4.37	1

8	4.5178	4.5106	1	8	4.36	4.37	1
8	4.4764	4.4679	1	8	4.36	4.38	1
9	4.41	4.41	1	9	4.38	4.38	1
9	4.33	4.41	1	9	4.37	4.37	1
9	4.38	4.36	1	9	4.41	4.41	1
9	4.41	4.4	1	9	4.39	4.38	1
9	4.34	4.42	1	9	4.38	4.38	1
9	4.35	4.37	1	9	4.41	4.41	1
10	4.432	4.429	1	10	4.62	4.63	1
10	4.443	4.417	1	10	4.65	4.62	1
10	4.43	4.417	1	10	4.62	4.63	1
10	4.443	4.445	1	10	4.56	4.6	1
10	4.404	4.413	1	10	4.61	4.62	1
10	4.427	4.416	1	10	4.61	4.62	1
11	4.71	4.88	1	11	4.3957	4.3957	1
11	4.66	4.67	1	11	4.3839	4.396	1
11	4.65	4.65	1	11	4.414	4.4198	1
11	4.94	4.96	1	11	4.408	4.4136	1
11	4.83	4.84	1	11	4.3993	4.4001	1
11	4.89	4.88	1	11	4.3756	4.3817	1
1	5.249	5.263	2	12	4.71	4.675	1
1	5.247	5.26	2	12	4.745	4.69	1
1	5.271	5.273	2	12	4.77	4.645	1
1	5.228	5.242	2	12	4.705	4.67	1
1	5.277	5.276	2	12	4.75	4.685	1
1	5.302	5.309	2	12	4.77	4.65	1
2	5.453	5.44	2	1	5.12	5.12	2
2	5.543	5.506	2	1	5.12	5.13	2
2	5.484	5.482	2	1	5.16	5.16	2
2	5.448	5.437	2	1	5.12	5.12	2
2	5.465	5.424	2	1	5.14	5.16	2
2	5.399	5.401	2	1	5.12	5.17	2
3	5.36	5.37	2	2	5.39	5.39	2
3	5.32	5.34	2	2	5.36	5.36	2
3	5.36	5.37	2	2	5.36	5.38	2
3	5.39	5.36	2	2	5.34	5.34	2
3	5.35	5.39	2	2	5.34	5.35	2
3	5.38	5.29	2	2	5.32	5.32	2
4	5.3279	5.3046	2	3	5.25	5.22	2
4	5.3006	5.3045	2	3	5.12	5.1	2
4	5.3034	5.2869	2	3	5.14	5.15	2
4	5.2973	5.3081	2	3	5.14	5.12	2
4	5.3024	5.2847	2	3	5.21	5.24	2
4	5.2847	5.2994	2	3	5.22	5.25	2
5	5.365	5.358	2	4	5.19	5.18	2
5	5.336	5.317	2	4	5.2	5.2	2
5	5.389	5.354	2	4	5.14	5.18	2
5	5.378	5.349	2	4	5.2	5.18	2
5	5.401	5.339	2	4	5.13	5.17	2
5	5.346	5.333	2	4	5.18	5.16	2

6	5.35	5.32	2	5	5.09	5.1	2
6	5.32	5.33	2	5	5.08	5.09	2
6	5.33	5.36	2	5	5.07	5.07	2
6	5.31	5.34	2	5	5.14	5.15	2
6	5.32	5.33	2	5	5.06	5.06	2
6	5.32	5.31	2	5	5.03	5.03	2
7	5.22	5.2	2	6	5.03	5.04	2
7	5.28	5.24	2	6	5.03	5.05	2
7	5.28	5.23	2	6	5.07	5.05	2
7	5.24	5.2	2	6	5.02	5.02	2
7	5.27	5.27	2	6	5.05	5.05	2
7	5.26	5.26	2	6	5.04	5.03	2
8	5.3525	5.3681	2	7	5.03	5.02	2
8	5.385	5.38	2	7	5.01	5.02	2
8	5.3823	5.3942	2	7	5.03	5.03	2
8	5.4065	5.386	2	7	5.07	5.06	2
8	5.4053	5.4135	2	7	5.02	5.02	2
8	5.4161	5.4155	2	7	5.04	5.04	2
9	5.32	5.28	2	8	5.02	5.04	2
9	5.25	5.28	2	8	5.04	5.07	2
9	5.35	5.35	2	8	5.05	5.04	2
9	5.32	5.28	2	8	4.97	4.98	2
9	5.28	5.25	2	8	4.97	4.97	2
9	5.36	5.32	2	8	4.9	4.98	2
10	5.314	5.332	2	9	5.04	5.05	2
10	5.338	5.319	2	9	5.03	5.04	2
10	5.323	5.302	2	9	5.06	5.06	2
10	5.305	5.321	2	9	5.05	5.05	2
10	5.319	5.331	2	9	5.04	5.03	2
10	5.303	5.327	2	9	5.06	5.06	2
11	5.57	5.58	2	10	5.09	5.08	2
11	5.56	5.79	2	10	5.07	5.08	2
11	5.61	5.61	2	10	5.09	5.11	2
11	6.11	6.09	2	10	5.08	5.1	2
11	5.71	5.74	2	10	5.05	5.05	2
11	5.73	5.74	2	10	5.06	5.08	2
				11	5.2413	5.2395	2
				11	5.1843	5.1853	2
				11	5.178	5.1806	2
				11	5.1374	5.1391	2
				11	5.1405	5.143	2
				11	5.1758	5.1744	2
				12	5.105	5.12	2
				12	5.09	5.165	2
				12	5.055	5.11	2
				12	5.0958	5.13	2
				12	5.1	5.17	2
				12	5.045	5.105	2

Fe₂O₃

Fe2O3: Glass Preparation				Fe2O3: Powder Preparation			
Lab	Duplicate 1	Duplicate 2	Cement	Lab	Duplicate 1	Duplicate 2	Cement
1	2.86	2.878	1	1	2.92	2.92	1
1	2.871	2.874	1	1	2.92	2.92	1
1	2.859	2.869	1	1	2.91	2.9	1
1	2.86	2.866	1	1	2.92	2.9	1
1	2.869	2.88	1	1	2.9	2.91	1
1	2.876	2.888	1	1	2.9	2.91	1
2	2.92	2.92	1	2	2.91	2.9	1
2	2.911	2.916	1	2	2.89	2.92	1
2	2.923	2.93	1	2	2.92	2.91	1
2	2.907	2.902	1	2	2.85	2.87	1
2	2.89	2.891	1	2	2.77	2.79	1
2	2.929	2.913	1	2	2.85	2.86	1
3	2.84	2.82	1	3	2.9	2.9	1
3	2.83	2.85	1	3	2.91	2.92	1
3	2.81	2.83	1	3	2.92	2.92	1
3	2.82	2.82	1	3	2.93	2.92	1
3	2.83	2.83	1	3	2.93	2.91	1
3	2.83	2.85	1	3	2.93	2.91	1
4	2.8786	2.8757	1	4	2.94	2.93	1
4	2.8757	2.8805	1	4	2.91	2.92	1
4	2.8746	2.8708	1	4	2.91	2.91	1
4	2.8862	2.8813	1	4	2.96	2.96	1
4	2.8785	2.8854	1	4	2.91	2.91	1
4	2.8833	2.8834	1	4	2.91	2.92	1
5	2.836	2.84	1	5	2.88	2.88	1
5	2.832	2.849	1	5	2.88	2.88	1
5	2.84	2.823	1	5	2.88	2.89	1
5	2.837	2.835	1	5	2.86	2.86	1
5	2.849	2.842	1	5	2.88	2.87	1
5	2.83	2.827	1	5	2.88	2.88	1
6	2.84	2.84	1	6	2.9	2.9	1
6	2.82	2.81	1	6	2.89	2.89	1
6	2.83	2.83	1	6	2.88	2.88	1
6	2.84	2.84	1	6	2.91	2.9	1
6	2.83	2.82	1	6	2.89	2.9	1
6	2.83	2.83	1	6	2.91	2.92	1
7	2.9	2.9	1	7	2.92	2.93	1
7	2.9	2.9	1	7	2.92	2.93	1
7	2.89	2.89	1	7	2.93	2.93	1
7	2.9	2.9	1	7	2.93	2.93	1
7	2.9	2.9	1	7	2.93	2.93	1
7	2.9	2.9	1	7	2.94	2.93	1
8	2.9013	2.9014	1	8	2.94	2.94	1
8	2.9014	2.9019	1	8	2.94	2.94	1
8	2.8897	2.891	1	8	2.93	2.95	1
8	2.8947	2.8945	1	8	2.95	2.95	1

8	2.8851	2.8857	1	8	2.94	2.93	1
8	2.8893	2.889	1	8	2.92	2.93	1
9	2.97	2.95	1	9	3.025	3.01	1
9	2.97	2.98	1	9	3.02	3.01	1
9	2.97	2.96	1	9	3.03	3.005	1
9	2.96	2.94	1	9	3.02	3.005	1
9	2.97	2.97	1	9	3.025	3.015	1
9	2.97	2.95	1	9	3.03	3.015	1
10	2.874	2.883	1	1	2.38	2.4	2
10	2.867	2.879	1	1	2.39	2.38	2
10	2.874	2.886	1	1	2.38	2.39	2
10	2.876	2.875	1	1	2.4	2.41	2
10	2.877	2.875	1	1	2.38	2.39	2
10	2.87	2.87	1	1	2.39	2.38	2
1	2.363	2.371	2	2	2.32	2.35	2
1	2.365	2.378	2	2	2.33	2.35	2
1	2.369	2.377	2	2	2.34	2.34	2
1	2.349	2.367	2	2	2.3	2.31	2
1	2.374	2.377	2	2	2.32	2.32	2
1	2.383	2.387	2	2	2.32	2.31	2
2	2.375	2.384	2	3	2.41	2.42	2
2	2.385	2.381	2	3	2.4	2.41	2
2	2.389	2.387	2	3	2.39	2.39	2
2	2.408	2.402	2	3	2.43	2.42	2
2	2.397	2.406	2	3	2.4	2.4	2
2	2.384	2.379	2	3	2.39	2.39	2
3	2.33	2.35	2	4	2.46	2.46	2
3	2.3	2.32	2	4	2.44	2.45	2
3	2.31	2.29	2	4	2.45	2.45	2
3	2.31	2.29	2	4	2.46	2.46	2
3	2.3	2.31	2	4	2.45	2.44	2
3	2.31	2.3	2	4	2.43	2.43	2
4	2.3734	2.3705	2	5	2.32	2.3	2
4	2.3714	2.3695	2	5	2.3	2.3	2
4	2.3694	2.3694	2	5	2.3	2.3	2
4	2.376	2.377	2	5	2.3	2.3	2
4	2.3761	2.377	2	5	2.3	2.3	2
4	2.3751	2.3799	2	5	2.3	2.3	2
5	2.348	2.347	2	6	2.35	2.35	2
5	2.353	2.343	2	6	2.34	2.34	2
5	2.346	2.354	2	6	2.35	2.35	2
5	2.352	2.327	2	6	2.36	2.36	2
5	2.35	2.343	2	6	2.35	2.36	2
5	2.336	2.329	2	6	2.36	2.37	2
6	2.33	2.33	2	7	2.36	2.36	2
6	2.32	2.32	2	7	2.38	2.38	2
6	2.35	2.35	2	7	2.37	2.38	2
6	2.33	2.33	2	7	2.36	2.36	2
6	2.32	2.31	2	7	2.38	2.38	2
6	2.35	2.35	2	7	2.38	2.38	2

7	2.39	2.39	2	8	2.39	2.39	2
7	2.39	2.39	2	8	2.37	2.37	2
7	2.39	2.39	2	8	2.38	2.37	2
7	2.4	2.4	2	8	2.38	2.39	2
7	2.38	2.39	2	8	2.38	2.37	2
7	2.39	2.4	2	8	2.36	2.37	2
8	2.4011	2.4026	2	9	2.46	2.45	2
8	2.3876	2.3878	2	9	2.46	2.48	2
8	2.3809	2.3804	2	9	2.46	2.47	2
8	2.3841	2.3853	2	9	2.46	2.45	2
8	2.3825	2.3826	2	9	2.45	2.475	2
8	2.3872	2.385	2	9	2.455	2.47	2
9	2.46	2.46	2				
9	2.47	2.47	2				
9	2.46	2.45	2				
9	2.46	2.45	2				
9	2.46	2.46	2				
9	2.45	2.45	2				
10	2.37	2.366	2				
10	2.371	2.361	2				
10	2.354	2.37	2				
10	2.373	2.36	2				
10	2.375	2.382	2				
10	2.364	2.38	2				

CaO: Glass Preparation				CaO: Powder Preparation			
Lab	Duplicate 1	Duplicate 2	Cement	Lab	Duplicate 1	Duplicate 2	Cement
3	62.54	62.57	1	1	62.8	62.87	1
3	62.44	62.53	1	1	62.68	62.76	1
3	62.58	62.63	1	1	62.81	62.84	1
3	62.39	62.48	1	1	62.82	62.66	1
3	62.47	62.45	1	1	62.76	62.7	1
3	62.6	62.62	1	1	62.77	62.84	1
4	62.409	62.452	1	2	62.74	62.72	1
4	62.452	62.415	1	2	62.75	62.67	1
4	62.31	62.47	1	2	62.56	62.75	1
4	62.349	62.351	1	2	62.78	62.63	1
4	62.4	62.504	1	2	62.6	62.63	1
4	62.412	62.432	1	2	62.7	62.57	1
5	61.67	61.766	1	5	62.67	62.67	1
5	61.947	61.946	1	5	62.87	62.89	1
5	61.77	61.693	1	5	62.59	62.67	1
5	61.83	61.813	1	5	62.74	62.75	1
5	61.869	61.739	1	5	62.77	62.78	1
5	61.706	61.782	1	5	62.56	62.53	1
6	62.32	62.32	1	6	62.75	62.72	1
6	62.08	62.11	1	6	62.8	62.81	1
6	62.31	62.35	1	6	62.8	62.72	1
6	62.51	62.52	1	6	63.01	62.88	1
6	62.29	62.29	1	6	62.91	62.85	1
6	62.34	62.33	1	6	62.87	62.83	1
7	62.48	62.49	1	7	62.5	62.43	1
7	62.51	62.53	1	7	62.45	62.4	1
7	62.45	62.47	1	7	62.41	62.39	1
7	62.47	62.49	1	7	62.26	62.29	1
7	62.55	62.56	1	7	62.53	62.53	1
7	62.48	62.49	1	7	62.58	62.54	1
8	62.514	62.585	1	8	62.41	62.47	1
8	62.564	62.53	1	8	62.39	62.49	1
8	62.316	62.308	1	8	62.36	62.45	1
8	62.45	62.447	1	8	62.54	62.44	1
8	62.325	62.347	1	8	62.67	62.65	1
8	62.353	62.349	1	8	62.67	62.69	1
9	63.28	62.99	1	9	62.61	62.61	1
9	63.29	63.47	1	9	62.54	63.53	1
9	63.26	63.14	1	9	62.53	62.53	1
9	63.4	62.95	1	9	62.59	62.62	1
9	63.4	63.54	1	9	62.55	62.56	1
9	63.31	63.12	1	9	62.55	62.54	1
10	62.331	62.328	1	10	62.73	62.78	1
10	62.38	62.325	1	10	62.78	62.89	1
10	62.321	62.315	1	10	62.92	62.94	1
10	62.326	62.341	1	10	62.55	62.7	1

10	62.42	62.381	1	10	62.69	62.69	1
10	62.392	62.327	1	10	62.69	62.84	1
3	63.12	63.15	2	11	62.847	62.93	1
3	63.12	63.13	2	11	62.897	62.938	1
3	63.13	63.16	2	11	63.194	63.175	1
3	62.9	62.86	2	11	62.926	62.968	1
3	62.89	63.01	2	11	62.993	62.993	1
3	63	62.99	2	11	62.8	62.745	1
4	63.4	63.42	2	1	63.97	64.16	2
4	63.382	63.401	2	1	64	64.01	2
4	63.391	63.372	2	1	64.04	63.98	2
4	63.349	63.372	2	1	63.91	64.05	2
4	63.456	63.39	2	1	64.01	64.05	2
4	63.393	63.397	2	1	64	64.09	2
5	62.859	62.937	2	2	63.75	63.83	2
5	62.836	62.883	2	2	63.74	63.84	2
5	63.014	62.994	2	2	63.95	63.77	2
5	63.066	62.904	2	2	63.72	63.84	2
5	63.035	63.081	2	2	63.63	63.78	2
5	63.022	62.953	2	2	63.67	63.65	2
6	63.37	63.4	2	5	63.81	63.82	2
6	63.27	63.28	2	5	64.15	64.17	2
6	63.41	63.42	2	5	63.93	63.95	2
6	63.42	63.43	2	5	63.75	63.77	2
6	63.47	63.44	2	5	63.63	63.65	2
6	63.44	63.41	2	5	63.85	63.86	2
7	63.34	63.38	2	6	63.31	63.31	2
7	63.27	63.29	2	6	63.28	63.34	2
7	63.25	63.26	2	6	63.41	63.29	2
7	63.38	63.37	2	6	63.29	63.33	2
7	63.29	63.28	2	6	63.29	63.33	2
7	63.29	63.25	2	6	63.25	63.25	2
8	63.283	63.292	2	7	63.67	63.57	2
8	63.333	63.327	2	7	63.44	63.41	2
8	63.089	63.1	2	7	63.58	63.59	2
8	63.224	63.238	2	7	63.69	63.65	2
8	63.151	63.162	2	7	63.27	63.24	2
8	63.262	63.245	2	7	63.74	63.71	2
9	63.73	63.78	2	8	63.39	63.43	2
9	63.87	64.07	2	8	63.32	63.37	2
9	63.64	63.65	2	8	63.5	63.59	2
9	63.59	63.77	2	8	63.78	63.87	2
9	63.93	64.05	2	8	63.7	63.79	2
9	63.59	63.62	2	8	63.6	63.54	2
10	63.322	63.224	2	9	63.3	63.27	2
10	63.314	63.291	2	9	63.17	63.16	2
10	63.308	63.255	2	9	63.26	63.25	2
10	63.301	63.315	2	9	63.29	63.31	2
10	63.237	63.259	2	9	63.16	63.17	2
10	63.237	63.265	2	9	63.28	63.28	2

10	64.28	64.31	2
10	64.25	64.24	2
10	64.33	64.27	2
10	64.18	64.23	2
10	64.07	64.12	2
10	64.12	64.13	2
11	64.203	64.226	2
11	64.022	64.001	2
11	63.975	63.93	2
11	63.66	63.678	2
11	63.662	63.736	2
11	63.91	63.883	2

MgO

MgO: Glass Preparation				MgO: Powder Preparation			
Lab	Duplicate 1	Duplicate 2	Cement	Lab	Duplicate 1	Duplicate 2	Cement
1	2.517	2.523	1	1	2.71	2.65	1
1	2.544	2.55	1	1	2.67	2.61	1
1	2.541	2.548	1	1	2.61	2.6	1
1	2.546	2.544	1	1	2.61	2.62	1
1	2.546	2.552	1	1	2.64	2.65	1
1	2.559	2.561	1	1	2.65	2.64	1
2	2.56	2.55	1	2	2.56	2.56	1
2	2.56	2.57	1	2	2.58	2.58	1
2	2.56	2.56	1	2	2.57	2.58	1
2	2.56	2.58	1	2	2.58	2.57	1
2	2.58	2.58	1	2	2.57	2.57	1
2	2.59	2.57	1	2	2.56	2.56	1
3	2.5776	2.5698	1	3	2.68	2.68	1
3	2.5689	2.5708	1	3	2.66	2.69	1
3	2.5678	2.563	1	3	2.7	2.72	1
3	2.5678	2.5916	1	3	2.69	2.7	1
3	2.5688	2.5631	1	3	2.6	2.63	1
3	2.56	2.561	1	3	2.64	2.66	1
4	2.55	2.56	1	4	2.65	2.64	1
4	2.55	2.54	1	4	2.63	2.63	1
4	2.56	2.55	1	4	2.64	2.64	1
4	2.54	2.55	1	4	2.66	2.66	1
4	2.55	2.55	1	4	2.63	2.64	1
4	2.54	2.54	1	4	2.64	2.64	1
5	2.56	2.58	1	5	2.56	2.56	1
5	2.57	2.57	1	5	2.6	2.59	1
5	2.57	2.57	1	5	2.59	2.59	1
5	2.59	2.59	1	5	2.6	2.62	1
5	2.59	2.58	1	5	2.59	2.59	1
5	2.59	2.59	1	5	2.61	2.61	1
6	2.56	2.55	1	6	2.61	2.61	1
6	2.55	2.54	1	6	2.62	2.63	1
6	2.57	2.53	1	6	2.61	2.62	1
6	2.56	2.55	1	6	2.57	2.58	1
6	2.55	2.54	1	6	2.61	2.61	1
6	2.56	2.56	1	6	2.58	2.58	1
7	2.5675	2.5773	1	7	2.66	2.65	1
7	2.6158	2.564	1	7	2.72	2.72	1
7	2.5894	2.5557	1	7	2.73	2.72	1
7	2.5683	2.6061	1	7	2.66	2.67	1
7	2.5666	2.5859	1	7	2.73	2.72	1
7	2.5389	2.5645	1	7	2.74	2.73	1
8	2.55	2.554	1	8	2.62	2.64	1
8	2.542	2.543	1	8	2.63	2.64	1
8	2.55	2.541	1	8	2.61	2.61	1
8	2.545	2.543	1	8	2.64	2.64	1

8	2.545	2.558	1	8	2.65	2.66	1
8	2.549	2.565	1	8	2.66	2.66	1
9	2.58	2.59	1	9	2.72	2.72	1
9	2.58	2.6	1	9	2.72	2.71	1
9	2.6	2.59	1	9	2.7	2.68	1
9	2.59	2.6	1	9	2.7	2.71	1
9	2.61	2.61	1	9	2.69	2.7	1
9	2.61	2.6	1	9	2.67	2.68	1
1	2.109	2.113	2	10	2.75	2.75	1
1	2.111	2.121	2	10	2.77	2.78	1
1	2.115	2.117	2	10	2.76	2.78	1
1	2.103	2.114	2	10	2.67	2.67	1
1	2.125	2.125	2	10	2.75	2.75	1
1	2.103	2.111	2	10	2.75	2.75	1
2	2.13	2.14	2	11	2.6135	2.6048	1
2	2.11	2.13	2	11	2.6256	2.6272	1
2	2.11	2.12	2	11	2.6696	2.6756	1
2	2.12	2.12	2	11	2.5855	2.5833	1
2	2.12	2.11	2	11	2.6288	2.6262	1
2	2.1	2.1	2	11	2.6164	2.6247	1
3	2.1086	2.1047	2	12	2.695	2.67	1
3	2.1086	2.1144	2	12	2.7	2.66	1
3	2.1007	2.1066	2	12	2.725	2.66	1
3	2.0975	2.1014	2	12	2.69	2.67	1
3	2.1083	2.0975	2	12	2.705	2.655	1
3	2.1092	2.0985	2	12	2.72	2.66	1
4	2.09	2.09	2	1	2.02	2.08	2
4	2.09	2.09	2	1	2.05	2.06	2
4	2.1	2.09	2	1	2.07	2.04	2
4	2.09	2.09	2	1	2.07	2.07	2
4	2.09	2.09	2	1	2.01	2.06	2
4	2.08	2.09	2	1	2.06	2.08	2
5	2.13	2.13	2	2	2.13	2.13	2
5	2.14	2.12	2	2	2.1	2.1	2
5	2.13	2.12	2	2	2.11	2.11	2
5	2.15	2.14	2	2	2.11	2.11	2
5	2.13	2.15	2	2	2.1	2.1	2
5	2.13	2.14	2	2	2.09	2.08	2
6	2.08	2.08	2	3	2.05	2.06	2
6	2.11	2.11	2	3	2.03	2.01	2
6	2.11	2.11	2	3	2.05	2.06	2
6	2.09	2.1	2	3	2.05	2.06	2
6	2.09	2.1	2	3	2.07	2.07	2
6	2.08	2.1	2	3	2.04	2.06	2
7	2.1304	2.1274	2	4	2.07	2.05	2
7	2.1151	2.0926	2	4	2.03	2.04	2
7	2.1286	2.1148	2	4	2.03	2.04	2
7	2.1277	2.1235	2	4	2.05	2.06	2
7	2.1341	2.1109	2	4	2.05	2.05	2
7	2.1156	2.1174	2	4	2.05	2.04	2

8	2.077	2.101	2	5	2.04	2.04	2
8	2.095	2.104	2	5	2.04	2.01	2
8	2.1	2.088	2	5	2.05	2.05	2
8	2.092	2.092	2	5	2.06	2.07	2
8	2.089	2.095	2	5	2.05	2.05	2
8	2.082	2.085	2	5	2.04	2.05	2
9	2.11	2.11	2	6	2.06	2.06	2
9	2.11	2.13	2	6	2.08	2.08	2
9	2.11	2.11	2	6	2.06	2.06	2
9	2.14	2.14	2	6	2.08	2.08	2
9	2.13	2.13	2	6	2.05	2.05	2
9	2.11	2.11	2	6	2.05	2.05	2
				7	2.07	2.06	2
				7	2.04	2.04	2
				7	2.05	2.04	2
				7	2.07	2.07	2
				7	2.03	2.04	2
				7	2.04	2.04	2
				8	2.04	2.05	2
				8	2.04	2.03	2
				8	2.03	2.04	2
				8	2.07	2.08	2
				8	2.05	2.06	2
				8	2.08	2.08	2
				9	2.06	2.06	2
				9	2.09	2.08	2
				9	2.09	2.08	2
				9	2.06	2.06	2
				9	2.08	2.08	2
				9	2.09	2.09	2
				10	2.1	2.11	2
				10	2.1	2.11	2
				10	2.11	2.1	2
				10	2.12	2.13	2
				10	2.1	2.09	2
				10	2.1	2.1	2
				11	2.1171	2.1211	2
				11	2.102	2.0983	2
				11	2.1072	2.1037	2
				11	2.0809	2.0786	2
				11	2.0761	2.0826	2
				11	2.0853	2.0957	2
				12	2.01	2.02	2
				12	2.01	2.045	2
				12	1.995	2	2
				12	2.01	2.02	2
				12	2.005	2.04	2
				12	1.995	2	2

118

SO3: Glass Preparation				SO3: Powder Preparation			
Lab	Duplicate 1	Duplicate 2	Cement	Lab	Duplicate 1	Duplicate 2	Cement
1	3.1930	3.2030	1	1	3.24	3.24	1
1	3.2280	3.2330	1	1	3.21	3.23	1
1	3.2350	3.2410	1	1	3.24	3.24	1
1	3.2550	3.2640	1	1	3.23	3.25	1
1	3.2370	3.2430	1	1	3.22	3.22	1
1	3.2430	3.2500	1	1	3.2	3.21	1
2	3.2580	3.2470	1	2	2.98	2.98	1
2	3.2480	3.2120	1	2	2.99	2.99	1
2	3.2730	3.2420	1	2	2.99	2.98	1
2	3.2260	3.2220	1	2	2.98	2.98	1
2	3.2290	3.2040	1	2	2.96	2.96	1
2	3.2360	3.2400	1	2	2.98	2.98	1
3	3.2900	3.3000	1	3	3.15	3.14	1
3	3.2700	3.2700	1	3	3.24	3.23	1
3	3.2800	3.2800	1	3	3.22	3.21	1
3	3.2900	3.2900	1	3	3.22	3.2	1
3	3.2900	3.2800	1	3	3.21	3.21	1
3	3.3000	3.3100	1	3	3.23	3.22	1
4	3.2929	3.2939	1	4	3.19	3.18	1
4	3.2919	3.2850	1	4	3.2	3.19	1
4	3.2996	3.3026	1	4	3.16	3.18	1
4	3.2847	3.2945	1	4	3.18	3.19	1
4	3.2936	3.3025	1	4	3.17	3.18	1
4	3.3024	3.2975	1	4	3.17	3.18	1
5	3.2100	3.2100	1	5	3.31	3.31	1
5	3.2100	3.2000	1	5	3.3	3.3	1
5	3.2000	3.1900	1	5	3.29	3.3	1
5	3.2300	3.2200	1	5	3.25	3.26	1
5	3.2300	3.2400	1	5	3.31	3.32	1
5	3.2300	3.2300	1	5	3.26	3.26	1
6	3.2600	3.2700	1	6	3.16	3.17	1
6	3.2500	3.2500	1	6	3.27	3.27	1
6	3.2500	3.2600	1	6	3.26	3.27	1
6	3.2600	3.2500	1	6	3.19	3.15	1
6	3.2500	3.2500	1	6	3.29	3.28	1
6	3.2600	3.2400	1	6	3.27	3.26	1
7	3.1869	3.1839	1	7	3.09	3.11	1
7	3.1865	3.1885	1	7	3.08	3.08	1
7	3.1685	3.1683	1	7	3.09	3.1	1
7	3.1821	3.1825	1	7	3.1	3.1	1
7	3.1796	3.1842	1	7	3.11	3.11	1
7	3.1675	3.1658	1	7	3.11	3.11	1
8	3.1400	3.1300	1	8	3.22	3.22	1
8	3.1000	3.1200	1	8	3.2	3.21	1
8	3.1300	3.1400	1	8	3.22	3.23	1
8	3.1400	3.1400	1	8	3.23	3.23	1

8	3.1200	3.1200	1	8	3.22	3.23	1
8	3.1400	3.1400	1	8	3.24	3.24	1
9	3.2840	3.2730	1	9	3.27	3.28	1
9	3.2750	3.2590	1	9	3.28	3.27	1
9	3.2800	3.2840	1	9	3.27	3.27	1
9	3.2870	3.2790	1	9	3.26	3.25	1
9	3.2900	3.2790	1	9	3.25	3.27	1
9	3.2720	3.2740	1	9	3.27	3.26	1
1	3.6680	3.6770	2	10	3.1705	3.1515	1
1	3.6540	3.6720	2	10	3.1705	3.1515	1
1	3.6690	3.6720	2	10	3.1773	3.1776	1
1	3.6310	3.6480	2	10	3.1462	3.1449	1
1	3.8650	3.6950	2	10	3.1575	3.1558	1
1	3.6840	3.6950	2	10	3.1336	3.1345	1
2	3.6410	3.6540	2	11	3.155	3.005	1
2	3.6760	3.0670	2	11	3.13	3	1
2	3.6690	3.6690	2	11	3.15	3.005	1
2	3.7190	3.7390	2	11	3.155	3.005	1
2	3.7190	3.7030	2	11	3.13	3	1
2	3.6740	3.6570	2	11	3.115	3.01	1
3	3.6900	3.7200	2	1	3.71	3.72	2
3	3.7200	3.7000	2	1	3.71	3.72	2
3	3.7400	3.7300	2	1	3.68	3.71	2
3	3.7200	3.7200	2	1	3.7	3.71	2
3	3.7200	3.7200	2	1	3.71	3.7	2
3	3.7200	3.7200	2	1	3.71	3.72	2
4	3.7187	3.7110	2	2	3.33	3.33	2
4	3.7128	3.7294	2	2	3.43	3.43	2
4	3.7147	3.7118	2	2	3.38	3.39	2
4	3.7218	3.7198	2	2	3.37	3.37	2
4	3.7336	3.7247	2	2	3.38	3.38	2
4	3.7306	3.7199	2	2	3.45	3.45	2
5	3.6000	3.6000	2	3	3.63	3.66	2
5	3.6100	3.6000	2	3	3.7	3.69	2
5	3.6100	3.6000	2	3	3.75	3.76	2
5	3.6100	3.6200	2	3	3.68	3.65	2
5	3.6300	3.6200	2	3	3.67	3.66	2
5	3.6300	3.6400	2	3	3.69	3.71	2
6	3.7100	3.6900	2	4	3.67	3.67	2
6	3.7000	3.7100	2	4	3.67	3.69	2
6	3.7100	3.7100	2	4	3.71	3.72	2
6	3.6900	3.7000	2	4	3.65	3.65	2
6	3.7200	3.7000	2	4	3.68	3.71	2
6	3.7200	3.7100	2	4	3.7	3.69	2
7	3.7760	3.7759	2	5	3.86	3.86	2
7	3.7995	3.7964	2	5	3.76	3.76	2
7	3.7816	3.7816	2	5	3.86	3.86	2
7	3.7989	3.7967	2	5	3.76	3.77	2
7	3.7836	3.7915	2	5	3.76	3.76	2
7	3.7937	3.7966	2	5	3.84	3.84	2

8	3.5200	3.5400	2	6	3.82	3.83	2
8	3.5200	3.5200	2	6	3.83	3.84	2
8	3.4800	3.4900	2	6	3.83	3.83	2
8	3.5200	3.5500	2	6	3.87	3.87	2
8	3.5100	3.5300	2	6	3.84	3.84	2
8	3.4800	3.5000	2	6	3.83	3.84	2
9	3.7090	3.7170	2	7	3.72	3.71	2
9	3.7160	3.6940	2	7	3.69	3.69	2
9	3.7120	3.6960	2	7	3.73	3.73	2
9	3.6940	3.6910	2	7	3.76	3.75	2
9	3.6920	3.6880	2	7	3.72	3.72	2
9	3.7000	3.7020	2	7	3.71	3.71	2
				8	3.75	3.75	2
				8	3.75	3.75	2
				8	3.75	3.76	2
				8	3.74	3.75	2
				8	3.78	3.78	2
				8	3.74	3.75	2
				9	3.74	3.74	2
				9	3.76	3.76	2
				9	3.72	3.73	2
				9	3.79	3.79	2
				9	3.79	3.79	2
				9	3.78	3.77	2
				10	3.7037	3.7	2
				10	3.7154	3.7171	2
				10	3.7118	3.7114	2
				10	3.6639	3.6617	2
				10	3.6778	3.6755	2
				10	3.6851	3.6885	2
				11	3.61	3.66	2
				11	3.7	3.56	2
				11	3.705	3.695	2
				11	3.605	3.67	2
				11	3.7	3.565	2
				11	3.715	3.69	2

Na₂O

Na2O: Glass Preparation				Na2O: Powder Preparation			
Lab	Duplicate 1	Duplicate 2	Cement	Lab	Duplicate 1	Duplicate 2	Cement
1	0.156	0.157	1	1	0.163	0.162	1
1	0.161	0.159	1	1	0.16	0.159	1
1	0.157	0.159	1	1	0.157	0.166	1
1	0.158	0.161	1	1	0.157	0.162	1
1	0.156	0.157	1	1	0.161	0.163	1
1	0.16	0.163	1	1	0.162	0.162	1
2	0.193	0.183	1	2	0.14	0.14	1
2	0.191	0.185	1	2	0.14	0.14	1
2	0.184	0.189	1	2	0.15	0.15	1
2	0.18	0.173	1	2	0.14	0.14	1
2	0.17	0.173	1	2	0.14	0.14	1
2	0.183	0.18	1	2	0.15	0.14	1
3	0.13	0.15	1	3	0.16	0.16	1
3	0.13	0.14	1	3	0.13	0.12	1
3	0.13	0.14	1	3	0.13	0.13	1
3	0.13	0.14	1	3	0.14	0.15	1
3	0.14	0.14	1	3	0.14	0.14	1
3	0.15	0.15	1	3	0.14	0.14	1
4	0.1515	0.1554	1	4	0.17	0.17	1
4	0.1524	0.1515	1	4	0.16	0.17	1
4	0.1524	0.1534	1	4	0.17	0.17	1
4	0.1485	0.1524	1	4	0.16	0.16	1
4	0.1524	0.1524	1	4	0.16	0.16	1
4	0.1563	0.1543	1	4	0.16	0.16	1
5	0.1281	0.1331	1	5	0.15	0.16	1
5	0.1316	0.1276	1	5	0.16	0.16	1
5	0.138	0.1361	1	5	0.16	0.16	1
5	0.1238	0.1257	1	5	0.15	0.16	1
5	0.1285	0.1284	1	5	0.16	0.16	1
5	0.1293	0.128	1	5	0.16	0.16	1
6	0.172	0.175	1	6	0.19	0.2	1
6	0.17	0.172	1	6	0.19	0.19	1
6	0.167	0.166	1	6	0.19	0.2	1
6	0.168	0.174	1	6	0.16	0.16	1
6	0.173	0.171	1	6	0.16	0.17	1
6	0.171	0.171	1	6	0.17	0.17	1
7	0.143	0.136	1	7	0.144	0.156	1
7	0.141	0.144	1	7	0.154	0.151	1
7	0.132	0.142	1	7	0.152	0.153	1
7	0.147	0.148	1	7	0.151	0.15	1
7	0.142	0.149	1	7	0.147	0.153	1
7	0.141	0.136	1	7	0.149	0.156	1
8	0.16532	0.16305	1	8	0.17	0.17	1
8	0.14847	0.14848	1	8	0.17	0.18	1
8	0.15666	0.15591	1	8	0.18	0.17	1
8	0.14982	0.14861	1	8	0.18	0.18	1

8	0.14229	0.15104	1	8	0.18	0.18	1
8	0.14863	0.137	1	8	0.17	0.18	1
9	0.17	0.16	1	9	0.13181	0.1303	1
9	0.15	0.15	1	9	0.12829	0.1319	1
9	0.14	0.16	1	9	0.1293	0.13239	1
9	0.18	0.16	1	9	0.12894	0.12486	1
9	0.16	0.16	1	9	0.1272	0.12338	1
9	0.16	0.15	1	9	0.12524	0.12324	1
10	0.161	0.168	1	10	0.19	0.19	1
10	0.159	0.166	1	10	0.2	0.19	1
10	0.17	0.165	1	10	0.2	0.19	1
10	0.157	0.168	1	10	0.19	0.19	1
10	0.165	0.158	1	10	0.2	0.19	1
10	0.167	0.162	1	10	0.2	0.19	1
1	0.152	0.149	2	1	0.159	0.16	2
1	0.15	0.151	2	1	0.155	0.158	2
1	0.149	0.15	2	1	0.153	0.151	2
1	0.151	0.151	2	1	0.16	0.159	2
1	0.15	0.151	2	1	0.159	0.147	2
1	0.15	0.15	2	1	0.156	0.151	2
2	0.165	0.172	2	2	0.14	0.13	2
2	0.172	0.174	2	2	0.14	0.14	2
2	0.172	0.172	2	2	0.14	0.14	2
2	0.158	0.162	2	2	0.14	0.14	2
2	0.167	0.165	2	2	0.14	0.14	2
2	0.16	0.164	2	2	0.14	0.14	2
3	0.12	0.14	2	3	0.15	0.15	2
3	0.12	0.14	2	3	0.15	0.15	2
3	0.13	0.14	2	3	0.15	0.15	2
3	0.13	0.13	2	3	0.12	0.13	2
3	0.13	0.13	2	3	0.13	0.13	2
3	0.14	0.13	2	3	0.12	0.12	2
4	0.145	0.1421	2	4	0.17	0.16	2
4	0.1441	0.146	2	4	0.17	0.17	2
4	0.1431	0.1441	2	4	0.17	0.17	2
4	0.147	0.1422	2	4	0.17	0.16	2
4	0.1451	0.1451	2	4	0.17	0.16	2
4	0.1431	0.1451	2	4	0.17	0.17	2
5	0.1222	0.1268	2	5	0.15	0.15	2
5	0.1201	0.1231	2	5	0.16	0.15	2
5	0.1201	0.1262	2	5	0.15	0.16	2
5	0.1239	0.1203	2	5	0.16	0.16	2
5	0.1228	0.1237	2	5	0.15	0.15	2
5	0.1176	0.1207	2	5	0.16	0.15	2
6	0.166	0.164	2	6	0.18	0.19	2
6	0.158	0.159	2	6	0.17	0.16	2
6	0.158	0.161	2	6	0.18	0.17	2
6	0.165	0.162	2	6	0.18	0.18	2
6	0.161	0.16	2	6	0.18	0.17	2
6	0.165	0.163	2	6	0.17	0.17	2

7	0.128	0.132	2	7	0.144	0.148	2
7	0.145	0.142	2	7	0.145	0.143	2
7	0.131	0.131	2	7	0.15	0.149	2
7	0.137	0.143	2	7	0.146	0.15	2
7	0.132	0.131	2	7	0.142	0.155	2
7	0.13	0.129	2	7	0.146	0.148	2
8	0.1321	0.12917	2	8	0.18	0.18	2
8	0.14767	0.14105	2	8	0.18	0.18	2
8	0.157	0.16856	2	8	0.18	0.17	2
8	0.15734	0.15703	2	8	0.18	0.18	2
8	0.14441	0.15827	2	8	0.19	0.19	2
8	0.12748	0.14918	2	8	0.18	0.18	2
9	0.14	0.14	2	9	0.12855	0.1331	2
9	0.14	0.15	2	9	0.12978	0.12912	2
9	0.14	0.14	2	9	0.12115	0.12421	2
9	0.14	0.15	2	9	0.12243	0.12013	2
9	0.14	0.14	2	9	0.11872	0.12337	2
9	0.14	0.14	2	9	0.12637	0.12628	2
10	0.161	0.153	2	10	0.19	0.2	2
10	0.166	0.161	2	10	0.2	0.19	2
10	0.146	0.153	2	10	0.19	0.19	2
10	0.157	0.15	2	10	0.19	0.2	2
10	0.154	0.161	2	10	0.195	0.19	2
10	0.156	0.16	2	10	0.19	0.19	2

Lab	K2O: Glass Preparation Duplicate 1	Duplicate 2	Cement	Lab	K2O: Powder Preparation Duplicate 1	Duplicate 2	Cement
1	0.719	0.721	1	1	0.75	0.74	1
1	0.723	0.722	1	1	0.74	0.74	1
1	0.724	0.728	1	1	0.75	0.74	1
1	0.732	0.725	1	1	0.74	0.74	1
1	0.722	0.72	1	1	0.74	0.74	1
1	0.723	0.727	1	1	0.73	0.73	1
2	0.74	0.73	1	2	0.73	0.73	1
2	0.74	0.74	1	2	0.73	0.73	1
2	0.74	0.73	1	2	0.74	0.73	1
2	0.73	0.73	1	2	0.73	0.73	1
2	0.74	0.74	1	2	0.73	0.73	1
2	0.74	0.74	1	2	0.74	0.74	1
3	0.7387	0.7377	1	3	0.73	0.73	1
3	0.7407	0.7367	1	3	0.74	0.75	1
3	0.7377	0.7387	1	3	0.74	0.74	1
3	0.7335	0.7325	1	3	0.74	0.74	1
3	0.7335	0.7326	1	3	0.74	0.74	1
3	0.7345	0.7345	1	3	0.74	0.74	1
4	0.7235	0.7217	1	4	0.75	0.75	1
4	0.728	0.7315	1	4	0.74	0.75	1
4	0.721	0.7236	1	4	0.75	0.74	1
4	0.7243	0.7223	1	4	0.75	0.75	1
4	0.7299	0.7328	1	4	0.75	0.74	1
4	0.7267	0.7261	1	4	0.75	0.75	1
5	0.73	0.73	1	5	0.74	0.74	1
5	0.73	0.73	1	5	0.74	0.74	1
5	0.73	0.73	1	5	0.74	0.74	1
5	0.74	0.73	1	5	0.73	0.74	1
5	0.73	0.73	1	5	0.74	0.74	1
5	0.73	0.73	1	5	0.73	0.73	1
6	0.74	0.73	1	6	0.74	0.74	1
6	0.74	0.73	1	6	0.75	0.75	1
6	0.74	0.74	1	6	0.75	0.74	1
6	0.74	0.74	1	6	0.74	0.75	1
6	0.74	0.74	1	6	0.75	0.74	1
6	0.74	0.73	1	6	0.75	0.75	1
7	0.728	0.727	1	7	0.725	0.726	1
7	0.728	0.726	1	7	0.724	0.724	1
7	0.727	0.724	1	7	0.734	0.734	1
7	0.729	0.732	1	7	0.73	0.728	1
7	0.723	0.728	1	7	0.729	0.728	1
7	0.731	0.727	1	7	0.73	0.731	1
8	0.732	0.731	1	8	0.73	0.73	1
8	0.7352	0.7359	1	8	0.73	0.73	1
8	0.7321	0.7316	1	8	0.73	0.73	1
8	0.7336	0.7341	1	8	0.73	0.73	1

8	0.7306	0.731	1	8	0.73	0.73	1
8	0.7322	0.7321	1	8	0.73	0.73	1
9	0.73	0.73	1	9	0.74	0.75	1
9	0.73	0.73	1	9	0.75	0.75	1
9	0.73	0.73	1	9	0.75	0.75	1
9	0.73	0.73	1	9	0.73	0.73	1
9	0.73	0.73	1	9	0.75	0.75	1
9	0.73	0.73	1	9	0.75	0.75	1
10	0.724	0.72	1	10	0.7204	0.7211	1
10	0.717	0.724	1	10	0.7191	0.7202	1
10	0.724	0.724	1	10	0.7307	0.7312	1
10	0.72	0.721	1	10	0.7189	0.7189	1
10	0.724	0.725	1	10	0.7284	0.7284	1
10	0.722	0.717	1	10	0.725	0.7255	1
11	0.75	0.74	1	1	1.17	1.19	2
11	0.75	0.74	1	1	1.18	1.18	2
11	0.75	0.74	1	1	1.19	1.19	2
11	0.75	0.74	1	1	1.17	1.17	2
11	0.75	0.75	1	1	1.18	1.19	2
11	0.75	0.75	1	1	1.2	1.18	2
1	1.162	1.161	2	2	1.17	1.16	2
1	1.17	1.17	2	2	1.19	1.19	2
1	1.154	1.152	2	2	1.18	1.18	2
1	1.157	1.158	2	2	1.17	1.18	2
1	1.168	1.17	2	2	1.18	1.18	2
1	1.156	1.158	2	2	1.19	1.19	2
2	1.18	1.19	2	3	1.18	1.18	2
2	1.19	1.18	2	3	1.19	1.19	2
2	1.18	1.18	2	3	1.2	1.2	2
2	1.19	1.19	2	3	1.18	1.19	2
2	1.19	1.19	2	3	1.19	1.19	2
2	1.18	1.18	2	3	1.19	1.19	2
3	1.1818	1.1779	2	4	1.19	1.19	2
3	1.1828	1.1789	2	4	1.2	1.2	2
3	1.1818	1.1798	2	4	1.2	1.21	2
3	1.1753	1.1802	2	4	1.19	1.19	2
3	1.1764	1.1773	2	4	1.21	1.21	2
3	1.1783	1.1773	2	4	1.2	1.21	2
4	1.1688	1.167	2	5	1.19	1.19	2
4	1.1561	1.1588	2	5	1.18	1.18	2
4	1.1587	1.1653	2	5	1.2	1.2	2
4	1.1656	1.1644	2	5	1.18	1.18	2
4	1.1622	1.165	2	5	1.19	1.19	2
4	1.1702	1.1601	2	5	1.19	1.19	2
5	1.18	1.18	2	6	1.23	1.23	2
5	1.18	1.19	2	6	1.24	1.24	2
5	1.19	1.19	2	6	1.24	1.24	2
5	1.18	1.19	2	6	1.23	1.23	2
5	1.18	1.19	2	6	1.24	1.24	2
5	1.18	1.18	2	6	1.25	1.24	2

6	1.17	1.17	2	7	1.162	1.163	2
6	1.17	1.17	2	7	1.168	1.17	2
6	1.17	1.18	2	7	1.17	1.172	2
6	1.18	1.18	2	7	1.167	1.169	2
6	1.17	1.17	2	7	1.178	1.179	2
6	1.18	1.17	2	7	1.165	1.166	2
7	1.167	1.174	2	8	1.16	1.16	2
7	1.166	1.172	2	8	1.16	1.17	2
7	1.172	1.174	2	8	1.17	1.17	2
7	1.165	1.165	2	8	1.16	1.16	2
7	1.165	1.174	2	8	1.17	1.17	2
7	1.164	1.166	2	8	1.17	1.17	2
8	1.1534	1.1541	2	9	1.2	1.21	2
8	1.1629	1.1634	2	9	1.22	1.21	2
8	1.1622	1.1609	2	9	1.2	1.2	2
8	1.1667	1.1665	2	9	1.21	1.21	2
8	1.1615	1.1623	2	9	1.21	1.21	2
8	1.1655	1.1647	2	9	1.22	1.22	2
9	1.17	1.18	2	10	1.1781	1.1751	2
9	1.18	1.17	2	10	1.1826	1.128	2
9	1.16	1.17	2	10	1.1817	1.1817	2
9	1.17	1.18	2	10	1.1675	1.1682	2
9	1.17	1.17	2	10	1.1772	1.178	2
9	1.17	1.17	2	10	1.1806	1.1805	2
10	1.161	1.16	2				
10	1.16	1.158	2				
10	1.165	1.159	2				
10	1.158	1.157	2				
10	1.16	1.16	2				
10	1.157	1.157	2				
11	1.21	1.21	2				
11	1.21	1.21	2				
11	1.22	1.22	2				
11	1.21	1.22	2				
11	1.21	1.21	2				
11	1.22	1.22	2				

TiO$_2$

TiO2: Glass Preparation				TiO2: Powder Preparation			
Lab	Duplicate 1	Duplicate 2	Cement	Lab	Duplicate 1	Duplicate 2	Cement
1	0.223	0.229	1	1	0.23	0.23	1
1	0.229	0.231	1	1	0.23	0.23	1
1	0.229	0.225	1	1	0.23	0.23	1
1	0.228	0.225	1	1	0.23	0.23	1
1	0.227	0.227	1	1	0.22	0.23	1
1	0.232	0.229	1	1	0.23	0.23	1
2	0.23	0.236	1	2	0.23	0.23	1
2	0.235	0.22	1	2	0.23	0.23	1
2	0.24	0.229	1	2	0.23	0.23	1
2	0.233	0.222	1	2	0.23	0.23	1
2	0.225	0.232	1	2	0.23	0.23	1
2	0.231	0.226	1	2	0.23	0.23	1
3	0.24	0.24	1	3	0.24	0.24	1
3	0.24	0.23	1	3	0.24	0.24	1
3	0.24	0.24	1	3	0.24	0.24	1
3	0.23	0.24	1	3	0.24	0.24	1
3	0.23	0.24	1	3	0.24	0.24	1
3	0.23	0.24	1	3	0.24	0.24	1
4	0.2306	0.2286	1	4	0.23	0.23	1
4	0.2306	0.2306	1	4	0.23	0.23	1
4	0.2306	0.2296	1	4	0.23	0.23	1
4	0.2295	0.2286	1	4	0.23	0.23	1
4	0.2305	0.2276	1	4	0.23	0.23	1
4	0.2295	0.2305	1	4	0.23	0.23	1
5	0.2444	0.2388	1	5	0.23	0.23	1
5	0.2462	0.2445	1	5	0.22	0.22	1
5	0.2468	0.2452	1	5	0.22	0.22	1
5	0.247	0.2361	1	5	0.23	0.22	1
5	0.2477	0.2429	1	5	0.23	0.23	1
5	0.2495	0.2433	1	5	0.22	0.22	1
6	0.23	0.23	1	6	0.23	0.23	1
6	0.23	0.23	1	6	0.23	0.23	1
6	0.23	0.23	1	6	0.23	0.24	1
6	0.23	0.23	1	6	0.23	0.23	1
6	0.22	0.23	1	6	0.23	0.23	1
6	0.23	0.23	1	6	0.23	0.23	1
7	0.22	0.22	1	7	0.23	0.23	1
7	0.22	0.22	1	7	0.23	0.23	1
7	0.22	0.22	1	7	0.23	0.23	1
7	0.23	0.22	1	7	0.23	0.23	1
7	0.22	0.22	1	7	0.23	0.23	1
7	0.22	0.23	1	7	0.23	0.23	1
8	0.226	0.227	1	8	0.23	0.23	1
8	0.225	0.222	1	8	0.22	0.22	1
8	0.225	0.226	1	8	0.22	0.22	1
8	0.223	0.226	1	8	0.23	0.22	1

8	0.223	0.228	1	8	0.22	0.23	1
8	0.226	0.227	1	8	0.23	0.23	1
9	0.2228	0.2217	1	9	0.228	0.229	1
9	0.2203	0.2203	1	9	0.23	0.226	1
9	0.2205	0.223	1	9	0.229	0.227	1
9	0.2217	0.2236	1	9	0.232	0.229	1
9	0.2279	0.2204	1	9	0.231	0.23	1
9	0.2187	0.2213	1	9	0.233	0.228	1
10	0.24	0.24	1	10	0.23	0.24	1
10	0.24	0.24	1	10	0.23	0.24	1
10	0.24	0.23	1	10	0.23	0.23	1
10	0.23	0.23	1	10	0.24	0.23	1
10	0.23	0.23	1	10	0.24	0.24	1
10	0.23	0.23	1	10	0.23	0.23	1
11	0.2261	0.2335	1	11	0.2312	0.2203	1
11	0.2267	0.2341	1	11	0.2254	0.2266	1
11	0.2325	0.2374	1	11	0.2251	0.2272	1
11	0.2321	0.2342	1	11	0.2319	0.2283	1
11	0.2349	0.2416	1	11	0.2274	0.2264	1
11	0.2275	0.2236	1	11	0.2219	0.229	1
12	0.24	0.24	1	12	0.23	0.23	1
12	0.24	0.24	1	12	0.23	0.23	1
12	0.24	0.24	1	12	0.235	0.23	1
12	0.25	0.25	1	12	0.23	0.23	1
12	0.24	0.25	1	12	0.23	0.23	1
12	0.25	0.25	1	12	0.235	0.23	1
1	0.226	0.225	2	1	0.22	0.23	2
1	0.221	0.228	2	1	0.22	0.22	2
1	0.222	0.223	2	1	0.23	0.22	2
1	0.224	0.223	2	1	0.22	0.22	2
1	0.225	0.222	2	1	0.22	0.22	2
1	0.228	0.224	2	1	0.23	0.22	2
2	0.225	0.224	2	2	0.22	0.22	2
2	0.214	0.222	2	2	0.22	0.22	2
2	0.216	0.224	2	2	0.22	0.22	2
2	0.219	0.227	2	2	0.23	0.22	2
2	0.219	0.226	2	2	0.22	0.22	2
2	0.22	0.219	2	2	0.22	0.22	2
3	0.23	0.22	2	3	0.24	0.24	2
3	0.23	0.23	2	3	0.24	0.24	2
3	0.24	0.23	2	3	0.24	0.24	2
3	0.23	0.22	2	3	0.23	0.24	2
3	0.23	0.23	2	3	0.23	0.24	2
3	0.23	0.23	2	3	0.24	0.23	2
4	0.2239	0.2259	2	4	0.23	0.22	2
4	0.2258	0.2258	2	4	0.22	0.22	2
4	0.2239	0.2239	2	4	0.22	0.22	2
4	0.224	0.224	2	4	0.22	0.22	2
4	0.225	0.224	2	4	0.22	0.23	2
4	0.2249	0.2249	2	4	0.22	0.22	2

5	0.2382	0.2389	2	5	0.22	0.22	2
5	0.2378	0.2372	2	5	0.23	0.22	2
5	0.2379	0.238	2	5	0.22	0.22	2
5	0.2421	0.235	2	5	0.23	0.22	2
5	0.2404	0.2325	2	5	0.22	0.22	2
5	0.2411	0.2448	2	5	0.21	0.21	2
6	0.22	0.23	2	6	0.22	0.22	2
6	0.22	0.22	2	6	0.22	0.23	2
6	0.22	0.23	2	6	0.22	0.23	2
6	0.22	0.22	2	6	0.23	0.22	2
6	0.22	0.23	2	6	0.23	0.22	2
6	0.23	0.22	2	6	0.23	0.22	2
7	0.22	0.22	2	7	0.22	0.22	2
7	0.22	0.22	2	7	0.22	0.22	2
7	0.22	0.22	2	7	0.22	0.22	2
7	0.22	0.22	2	7	0.22	0.22	2
7	0.22	0.22	2	7	0.22	0.22	2
7	0.22	0.22	2	7	0.22	0.22	2
8	0.221	0.22	2	8	0.21	0.22	2
8	0.217	0.219	2	8	0.21	0.21	2
8	0.221	0.219	2	8	0.21	0.21	2
8	0.22	0.219	2	8	0.21	0.21	2
8	0.219	0.216	2	8	0.21	0.22	2
8	0.216	0.221	2	8	0.21	0.21	2
9	0.2191	0.2165	2	9	0.219	0.219	2
9	0.2146	0.217	2	9	0.218	0.222	2
9	0.2193	0.2147	2	9	0.22	0.218	2
9	0.2155	0.2184	2	9	0.221	0.219	2
9	0.2145	0.2157	2	9	0.219	0.222	2
9	0.2141	0.2158	2	9	0.221	0.22	2
10	0.24	0.24	2	10	0.23	0.21	2
10	0.24	0.23	2	10	0.22	0.22	2
10	0.24	0.24	2	10	0.22	0.23	2
10	0.23	0.23	2	10	0.22	0.22	2
10	0.23	0.23	2	10	0.22	0.23	2
10	0.23	0.23	2	10	0.21	0.22	2
11	0.221	0.2256	2	11	0.224	0.2183	2
11	0.2292	0.222	2	11	0.217	0.2176	2
11	0.2306	0.2269	2	11	0.2144	0.218	2
11	0.2183	0.2286	2	11	0.2171	0.2145	2
11	0.2206	0.2267	2	11	0.2204	0.213	2
11	0.2204	0.2257	2	11	0.2194	0.2198	2
12	0.24	0.24	2	12	0.22	0.22	2
12	0.24	0.24	2	12	0.22	0.22	2
12	0.24	0.24	2	12	0.22	0.22	2
12	0.24	0.25	2	12	0.22	0.22	2
12	0.24	0.24	2	12	0.22	0.22	2
12	0.24	0.24	2	12	0.22	0.22	2

P2O5: Glass Preparation				P2O5: Powder Preparation			
Lab	Duplicate 1	Duplicate 2	Cement	Lab	Duplicate 1	Duplicate 2	Cement
1	0.139	0.138	1	1	0.138	0.137	1
1	0.14	0.14	1	1	0.138	0.137	1
1	0.139	0.14	1	1	0.137	0.138	1
1	0.139	0.139	1	1	0.139	0.139	1
1	0.14	0.141	1	1	0.137	0.139	1
1	0.147	0.14	1	1	0.135	0.137	1
2	0.132	0.129	1	2	0.14	0.14	1
2	0.128	0.132	1	2	0.14	0.14	1
2	0.133	0.129	1	2	0.14	0.14	1
2	0.131	0.132	1	2	0.14	0.14	1
2	0.131	0.13	1	2	0.14	0.14	1
2	0.13	0.129	1	2	0.14	0.14	1
3	0.1368	0.1378	1	3	0.14	0.14	1
3	0.1358	0.1368	1	3	0.13	0.14	1
3	0.1378	0.1358	1	3	0.14	0.13	1
3	0.1377	0.1377	1	3	0.13	0.13	1
3	0.1377	0.1358	1	3	0.13	0.13	1
3	0.1358	0.1367	1	3	0.13	0.13	1
4	0.1339	0.1361	1	4	0.14	0.14	1
4	0.1377	0.14	1	4	0.15	0.15	1
4	0.139	0.1333	1	4	0.15	0.14	1
4	0.1339	0.139	1	4	0.14	0.14	1
4	0.1349	0.1378	1	4	0.15	0.15	1
4	0.1324	0.1379	1	4	0.14	0.14	1
5	0.14	0.14	1	5	0.14	0.14	1
5	0.14	0.14	1	5	0.14	0.14	1
5	0.14	0.14	1	5	0.14	0.14	1
5	0.14	0.14	1	5	0.14	0.14	1
5	0.14	0.14	1	5	0.14	0.13	1
5	0.14	0.14	1	5	0.14	0.14	1
6	0.136	0.138	1	6	0.139	0.133	1
6	0.137	0.136	1	6	0.139	0.133	1
6	0.136	0.137	1	6	0.139	0.131	1
6	0.137	0.137	1	6	0.139	0.136	1
6	0.136	0.138	1	6	0.136	0.141	1
6	0.138	0.137	1	6	0.136	0.138	1
7	0.141	0.136	1	7	0.14	0.14	1
7	0.138	0.14	1	7	0.14	0.14	1
7	0.138	0.139	1	7	0.14	0.14	1
7	0.141	0.139	1	7	0.14	0.14	1
7	0.141	0.141	1	7	0.14	0.14	1
7	0.138	0.141	1	7	0.14	0.14	1
8	0.13	0.13	1	8	0.1	0.1	1
8	0.13	0.13	1	8	0.1	0.1	1
8	0.13	0.13	1	8	0.1	0.1	1
8	0.13	0.13	1	8	0.1	0.1	1

8	0.13	0.13	1	8	0.1	0.1	1
8	0.13	0.13	1	8	0.1	0.1	1
9	0.14	0.1376	1	9	0.151	0.151	1
9	0.1381	0.1369	1	9	0.15	0.152	1
9	0.1303	0.1415	1	9	0.152	0.15	1
9	0.1371	0.1421	1	9	0.152	0.153	1
9	0.1405	0.1473	1	9	0.153	0.153	1
9	0.1344	0.1394	1	9	0.152	0.152	1
1	0.135	0.137	2	10	0.135	0.137	1
1	0.137	0.136	2	10	0.139	0.137	1
1	0.137	0.138	2	10	0.138	0.138	1
1	0.136	0.136	2	10	0.138	0.14	1
1	0.138	0.138	2	10	0.137	0.135	1
1	0.137	0.138	2	10	0.138	0.139	1
2	0.13	0.133	2	11	0.1446	0.1446	1
2	0.126	0.134	2	11	0.1394	0.1425	1
2	0.129	0.134	2	11	0.1367	0.1329	1
2	0.129	0.132	2	11	0.1504	0.1418	1
2	0.131	0.133	2	11	0.1396	0.14	1
2	0.135	0.125	2	11	0.1402	0.1337	1
3	0.1343	0.1343	2	1	0.137	0.133	2
3	0.1343	0.1343	2	1	0.136	0.138	2
3	0.1353	0.1353	2	1	0.137	0.137	2
3	0.1334	0.1334	2	1	0.137	0.137	2
3	0.1344	0.1344	2	1	0.134	0.137	2
3	0.1324	0.1344	2	1	0.137	0.137	2
4	0.1306	0.1337	2	2	0.14	0.14	2
4	0.1344	0.1293	2	2	0.14	0.14	2
4	0.1296	0.134	2	2	0.14	0.14	2
4	0.1347	0.136	2	2	0.14	0.14	2
4	0.1279	0.1325	2	2	0.14	0.14	2
4	0.1294	0.1357	2	2	0.14	0.14	2
5	0.14	0.14	2	3	0.13	0.13	2
5	0.14	0.14	2	3	0.13	0.13	2
5	0.14	0.14	2	3	0.13	0.13	2
5	0.14	0.14	2	3	0.14	0.14	2
5	0.14	0.14	2	3	0.14	0.14	2
5	0.14	0.14	2	3	0.14	0.13	2
6	0.134	0.133	2	4	0.14	0.14	2
6	0.135	0.134	2	4	0.14	0.15	2
6	0.135	0.135	2	4	0.14	0.14	2
6	0.135	0.136	2	4	0.15	0.14	2
6	0.135	0.136	2	4	0.14	0.14	2
6	0.134	0.136	2	4	0.14	0.14	2
7	0.137	0.136	2	5	0.14	0.14	2
7	0.135	0.14	2	5	0.14	0.14	2
7	0.141	0.14	2	5	0.14	0.14	2
7	0.138	0.137	2	5	0.13	0.14	2
7	0.138	0.135	2	5	0.14	0.14	2
7	0.133	0.134	2	5	0.14	0.14	2

8	0.13	0.13	2	6	0.142	0.14	2
8	0.13	0.13	2	6	0.147	0.152	2
8	0.13	0.13	2	6	0.141	0.143	2
8	0.13	0.13	2	6	0.136	0.14	2
8	0.13	0.13	2	6	0.135	0.138	2
8	0.13	0.13	2	6	0.142	0.137	2
9	0.1372	0.1369	2	7	0.14	0.14	2
9	0.1413	0.1334	2	7	0.14	0.14	2
9	0.1325	0.1372	2	7	0.14	0.15	2
9	0.1403	0.1395	2	7	0.14	0.14	2
9	0.1351	0.138	2	7	0.14	0.14	2
9	0.1361	0.1354	2	7	0.14	0.14	2
				8	0.11	0.11	2
				8	0.1	0.1	2
				8	0.11	0.11	2
				8	0.11	0.11	2
				8	0.1	0.11	2
				8	0.11	0.1	2
				9	0.154	0.155	2
				9	0.152	0.153	2
				9	0.154	0.153	2
				9	0.154	0.156	2
				9	0.155	0.155	2
				9	0.154	0.154	2
				10	0.143	0.14	2
				10	0.143	0.144	2
				10	0.14	0.14	2
				10	0.14	0.141	2
				10	0.141	0.142	2
				10	0.143	0.142	2
				11	0.1388	0.1407	2
				11	0.1396	0.1389	2
				11	0.1346	0.1352	2
				11	0.1407	0.141	2
				11	0.1369	0.1356	2
				11	0.1388	0.1374	2

Cl Glass and Powder				
Lab	Duplicate 1	Duplicate 2	Cement	Method
1	0.015	0.016	1	P
1	0.016	0.016	1	P
1	0.016	0.017	1	P
1	0.016	0.016	1	P
1	0.016	0.016	1	P
1	0.016	0.016	1	P
1	0.013	0.013	2	P
1	0.013	0.013	2	P
1	0.013	0.013	2	P
1	0.013	0.013	2	P
1	0.013	0.013	2	P
1	0.012	0.013	2	P
2	0.011	0.011	1	G
2	0.012	0.012	1	G
2	0.011	0.012	1	G
2	0.011	0.013	1	G
2	0.01	0.011	1	G
2	0.011	0.011	1	G
2	0.008	0.008	2	G
2	0.008	0.008	2	G
2	0.007	0.008	2	G
2	0.007	0.008	2	G
2	0.008	0.009	2	G
2	0.008	0.008	2	G
3	0.01	0.01	1	P
3	0.01	0.01	1	P
3	0.01	0.01	1	P
3	0.01	0.01	1	P
3	0.01	0.01	1	P
3	0.01	0.01	1	P
3	0.01	0	2	P
3	0.01	0.01	2	P
3	0.01	0.01	2	P
3	0.01	0.01	2	P
3	0	0.01	2	P
3	0	0	2	P
4	0.01	0.01	1	P
4	0.011	0.012	1	P
4	0.01	0.01	1	P
4	0.011	0.011	1	P
4	0.01	0.01	1	P
4	0.011	0.011	1	P
4	0.008	0.008	2	P
4	0.007	0.007	2	P
4	0.007	0.007	2	P
4	0.008	0.007	2	P

4	0.006	0.008	2	P
4	0.007	0.007	2	P
5	0.016	0.015	1	P
5	0.016	0.016	1	P
5	0.015	0.017	1	P
5	0.017	0.016	1	P
5	0.016	0.016	1	P
5	0.016	0.015	1	P
5	0.011	0.012	2	P
5	0.012	0.012	2	P
5	0.013	0.013	2	P
5	0.013	0.012	2	P
5	0.013	0.012	2	P
5	0.013	0.013	2	P
6	0.02743	0.02738	1	G
6	0.02741	0.02858	1	G
6	0.02616	0.02717	1	G
6	0.02644	0.02472	1	G
6	0.02748	0.0297	1	G
6	0.02572	0.02708	1	G
6	0.02109	0.02184	2	G
6	0.01991	0.02143	2	G
6	0.01985	0.02078	2	G
6	0.02024	0.02162	2	G
6	0.02116	0.02039	2	G
6	0.02258	0.02296	2	G

Mn2O3: Glass Preparation				Mn2O3: Powder Preparation			
Lab	Duplicate 1	Duplicate 2	Cement	Lab	Duplicate 1	Duplicate 2	Cement
1	0.203	0.204	1	1	0.206	0.206	1
1	0.206	0.206	1	1	0.202	0.206	1
1	0.206	0.205	1	1	0.199	0.203	1
1	0.206	0.204	1	1	0.207	0.203	1
1	0.207	0.208	1	1	0.204	0.205	1
1	0.207	0.207	1	1	0.202	0.203	1
2	0.21	0.205	1	2	0.21	0.21	1
2	0.21	0.205	1	2	0.21	0.21	1
2	0.208	0.208	1	2	0.21	0.21	1
2	0.209	0.207	1	2	0.21	0.21	1
2	0.209	0.209	1	2	0.21	0.21	1
2	0.217	0.206	1	2	0.21	0.21	1
3	0.2013	0.2013	1	3	0.21	0.21	1
3	0.2032	0.2023	1	3	0.21	0.21	1
3	0.2023	0.2013	1	3	0.21	0.21	1
3	0.2041	0.2032	1	3	0.21	0.21	1
3	0.2032	0.2041	1	3	0.21	0.21	1
3	0.2012	0.2022	1	3	0.21	0.21	1
4	0.203	0.203	1	4	0.206	0.206	1
4	0.201	0.201	1	4	0.207	0.21	1
4	0.202	0.202	1	4	0.209	0.207	1
4	0.203	0.204	1	4	0.207	0.211	1
4	0.203	0.201	1	4	0.209	0.208	1
4	0.203	0.202	1	4	0.206	0.207	1
5	0.203	0.206	1	5	0.2026	0.2017	1
5	0.2	0.202	1	5	0.2022	0.1975	1
5	0.202	0.203	1	5	0.203	0.1998	1
5	0.201	0.199	1	5	0.2026	0.206	1
5	0.199	0.199	1	5	0.2035	0.2056	1
5	0.202	0.199	1	5	0.2009	0.2031	1
6	0.2002	0.1976	1	6	0.183	0.187	1
6	0.1986	0.199	1	6	0.182	0.184	1
6	0.2033	0.2015	1	6	0.181	0.185	1
6	0.1995	0.1985	1	6	0.182	0.184	1
6	0.2014	0.2021	1	6	0.185	0.187	1
6	0.196	0.1956	1	6	0.185	0.186	1
7	0.19	0.19	1	7	0.21	0.2	1
7	0.19	0.19	1	7	0.21	0.21	1
7	0.19	0.19	1	7	0.22	0.21	1
7	0.19	0.19	1	7	0.21	0.2	1
7	0.19	0.19	1	7	0.21	0.21	1
7	0.19	0.19	1	7	0.21	0.21	1
1	0.051	0.053	2	1	0.059	0.056	2
1	0.053	0.054	2	1	0.054	0.055	2
1	0.052	0.054	2	1	0.053	0.055	2
1	0.052	0.053	2	1	0.058	0.056	2

1	0.053	0.053	2	1	0.055	0.055	2
1	0.055	0.054	2	1	0.056	0.056	2
2	0.059	0.059	2	2	0.06	0.06	2
2	0.06	0.053	2	2	0.06	0.06	2
2	0.058	0.058	2	2	0.06	0.06	2
2	0.056	0.055	2	2	0.06	0.06	2
2	0.055	0.056	2	2	0.06	0.06	2
2	0.056	0.054	2	2	0.06	0.06	2
3	0.0526	0.0535	2	3	0.06	0.06	2
3	0.0526	0.0535	2	3	0.05	0.05	2
3	0.0526	0.0535	2	3	0.05	0.05	2
3	0.0526	0.0545	2	3	0.06	0.05	2
3	0.0536	0.0536	2	3	0.05	0.05	2
3	0.0536	0.0536	2	3	0.05	0.05	2
4	0.056	0.055	2	4	0.06	0.057	2
4	0.055	0.056	2	4	0.06	0.061	2
4	0.055	0.055	2	4	0.057	0.057	2
4	0.056	0.055	2	4	0.057	0.058	2
4	0.055	0.055	2	4	0.058	0.057	2
4	0.054	0.055	2	4	0.057	0.056	2
5	0.052	0.054	2	5	0.0572	0.055	2
5	0.052	0.054	2	5	0.0532	0.0511	2
5	0.054	0.052	2	5	0.0541	0.0549	2
5	0.053	0.051	2	5	0.0545	0.0531	2
5	0.052	0.052	2	5	0.0524	0.0514	2
5	0.052	0.053	2	5	0.0548	0.0554	2
6	0.057	0.0627	2	6	0.05	0.049	2
6	0.0586	0.0558	2	6	0.05	0.051	2
6	0.0593	0.058	2	6	0.051	0.049	2
6	0.0598	0.0573	2	6	0.05	0.05	2
6	0.0589	0.0587	2	6	0.05	0.051	2
6	0.0534	0.0563	2	6	0.05	0.052	2
7	0.07	0.07	2	7	0.06	0.06	2
7	0.07	0.07	2	7	0.05	0.05	2
7	0.07	0.07	2	7	0.05	0.05	2
7	0.07	0.07	2	7	0.05	0.05	2
7	0.07	0.07	2	7	0.06	0.05	2
7	0.07	0.07	2	7	0.05	0.05	2

ZnO Lab	Duplicate 1	Duplicate 2	Cement	Method
2	0.014	0.014	1	G
2	0.014	0.014	1	G
2	0.013	0.014	1	G
2	0.013	0.013	1	G
2	0.013	0.014	1	G
2	0.014	0.014	1	G
3	0.0156	0.0156	1	G
3	0.0166	0.0166	1	G
3	0.0166	0.0166	1	G
3	0.0166	0.0166	1	G
3	0.0166	0.0166	1	G
3	0.0166	0.0166	1	G
6	0.01	0.01	1	G
6	0.01	0.01	1	G
6	0.01	0.01	1	G
6	0.01	0.01	1	G
6	0.01	0.01	1	G
6	0.01	0.01	1	G
8	0.011	0.011	1	G
8	0.01	0.01	1	G
8	0.01	0.011	1	G
8	0.011	0.011	1	G
8	0.01	0.01	1	G
8	0.011	0.011	1	G
9	0.01207	0.01227	1	G
9	0.01331	0.01306	1	G
9	0.01292	0.01376	1	G
9	0.01362	0.01287	1	G
9	0.01193	0.01153	1	G
9	0.01368	0.01246	1	G
10	0.014	0.0129	1	G
10	0.0137	0.0127	1	G
10	0.0144	0.0136	1	G
10	0.0151	0.0144	1	G
10	0.0137	0.013	1	G
10	0.0133	0.0129	1	G
1	0.014	0.014	1	P
1	0.014	0.014	1	P
1	0.014	0.014	1	P
1	0.014	0.014	1	P
1	0.014	0.014	1	P
1	0.014	0.014	1	P
4	0.01	0.01	1	P
4	0.01	0.01	1	P
4	0.01	0.01	1	P
4	0.01	0.01	1	P

4	0.01	0.01	1	P
4	0.01	0.01	1	P
5	0.013	0.014	1	P
5	0.013	0.013	1	P
5	0.013	0.013	1	P
5	0.013	0.013	1	P
5	0.013	0.013	1	P
5	0.013	0.013	1	P
7	0.0246	0.025	1	P
7	0.0236	0.0248	1	P
7	0.0246	0.0241	1	P
7	0.0254	0.0256	1	P
7	0.0248	0.0251	1	P
7	0.0255	0.0253	1	P
2	0.013	0.012	2	G
2	0.013	0.013	2	G
2	0.012	0.013	2	G
2	0.013	0.013	2	G
2	0.013	0.013	2	G
2	0.012	0.013	2	G
3	0.0156	0.0156	2	G
3	0.0156	0.0146	2	G
3	0.0146	0.0146	2	G
3	0.0156	0.0156	2	G
3	0.0156	0.0156	2	G
3	0.0156	0.0156	2	G
6	0.01	0.01	2	G
6	0.01	0.01	2	G
6	0.01	0.01	2	G
6	0.01	0.01	2	G
6	0.01	0.01	2	G
6	0.01	0.01	2	G
8	0.009	0.009	2	G
8	0.01	0.01	2	G
8	0.009	0.01	2	G
8	0.01	0.01	2	G
8	0.01	0.009	2	G
8	0.01	0.01	2	G
9	0.01316	0.01168	2	G
9	0.01214	0.01226	2	G
9	0.01218	0.01152	2	G
9	0.01031	0.01164	2	G
9	0.01203	0.01254	2	G
9	0.01256	0.01126	2	G
10	0.0131	0.0133	2	G
10	0.0124	0.0133	2	G
10	0.0125	0.0133	2	G
10	0.0115	0.0126	2	G
10	0.0117	0.013	2	G
10	0.0129	0.0115	2	G

139

1	0.014	0.014	2	P
1	0.013	0.013	2	P
1	0.013	0.013	2	P
1	0.013	0.013	2	P
1	0.013	0.013	2	P
1	0.013	0.013	2	P
4	0.01	0.01	2	P
4	0.01	0.01	2	P
4	0.01	0.01	2	P
4	0.01	0.01	2	P
4	0.01	0.01	2	P
4	0.01	0.01	2	P
5	0.012	0.012	2	P
5	0.012	0.012	2	P
5	0.012	0.012	2	P
5	0.012	0.012	2	P
5	0.012	0.012	2	P
5	0.012	0.012	2	P
7	0.0233	0.0233	2	P
7	0.0227	0.0238	2	P
7	0.0235	0.0224	2	P
7	0.0243	0.0241	2	P
7	0.0229	0.0237	2	P
7	0.0237	0.0242	2	P

SrO: Glass Preparation				SrO: Powder Preparation			
Lab	Duplicate 1	Duplicate 2	Cement	Lab	Duplicate 1	Duplicate 2	Cement
1	0.126	0.127	1	1	0.13	0.13	1
1	0.127	0.127	1	1	0.131	0.13	1
1	0.127	0.127	1	1	0.13	0.129	1
1	0.128	0.128	1	1	0.129	0.131	1
1	0.127	0.128	1	1	0.13	0.131	1
1	0.128	0.128	1	1	0.13	0.13	1
2	0.13	0.13	1	2	0.13	0.13	1
2	0.13	0.13	1	2	0.13	0.13	1
2	0.13	0.13	1	2	0.13	0.13	1
2	0.13	0.13	1	2	0.13	0.13	1
2	0.13	0.13	1	2	0.13	0.13	1
2	0.13	0.13	1	2	0.13	0.13	1
3	0.1309	0.1309	1	3	0.12	0.12	1
3	0.1309	0.1309	1	3	0.12	0.12	1
3	0.1309	0.1309	1	3	0.12	0.12	1
3	0.1299	0.1299	1	3	0.12	0.12	1
3	0.1309	0.1309	1	3	0.11	0.11	1
3	0.1299	0.1309	1	3	0.12	0.12	1
4	0.12	0.12	1	4	0.13	0.13	1
4	0.12	0.12	1	4	0.13	0.13	1
4	0.12	0.12	1	4	0.13	0.13	1
4	0.12	0.12	1	4	0.13	0.13	1
4	0.12	0.12	1	4	0.13	0.13	1
4	0.12	0.12	1	4	0.13	0.13	1
5	0.13	0.13	1	5	0.13	0.13	1
5	0.132	0.132	1	5	0.13	0.13	1
5	0.129	0.13	1	5	0.13	0.13	1
5	0.131	0.13	1	5	0.13	0.13	1
5	0.132	0.133	1	5	0.13	0.13	1
5	0.13	0.13	1	5	0.13	0.13	1
6	0.1263	0.1264	1	6	0.12	0.12	1
6	0.1264	0.1277	1	6	0.12	0.12	1
6	0.1278	0.1285	1	6	0.12	0.12	1
6	0.127	0.1288	1	6	0.12	0.12	1
6	0.1277	0.1262	1	6	0.12	0.12	1
6	0.127	0.1256	1	6	0.12	0.12	1
1	0.085	0.085	2	7	0.1296	0.1281	1
1	0.085	0.086	2	7	0.1293	0.1292	1
1	0.085	0.085	2	7	0.1289	0.1285	1
1	0.085	0.085	2	7	0.1291	0.1285	1
1	0.086	0.086	2	7	0.1284	0.1293	1
1	0.086	0.086	2	7	0.1282	0.1288	1
2	0.09	0.09	2	8	0.118	0.119	1
2	0.09	0.09	2	8	0.117	0.121	1
2	0.09	0.09	2	8	0.12	0.121	1
2	0.09	0.09	2	8	0.121	0.121	1

2	0.09	0.09	2	8	0.121	0.122	1
2	0.09	0.09	2	8	0.122	0.121	1
3	0.0886	0.0886	2	1	0.088	0.089	2
3	0.0886	0.0886	2	1	0.088	0.089	2
3	0.0886	0.0886	2	1	0.088	0.087	2
3	0.0876	0.0876	2	1	0.087	0.088	2
3	0.0886	0.0886	2	1	0.088	0.088	2
3	0.0886	0.0886	2	1	0.087	0.088	2
4	0.08	0.08	2	2	0.09	0.09	2
4	0.08	0.08	2	2	0.09	0.09	2
4	0.08	0.08	2	2	0.09	0.09	2
4	0.08	0.08	2	2	0.09	0.09	2
4	0.08	0.08	2	2	0.09	0.09	2
4	0.08	0.08	2	2	0.09	0.09	2
5	0.088	0.088	2	3	0.08	0.08	2
5	0.087	0.087	2	3	0.08	0.08	2
5	0.087	0.086	2	3	0.08	0.08	2
5	0.088	0.088	2	3	0.08	0.08	2
5	0.087	0.087	2	3	0.08	0.08	2
5	0.088	0.087	2	3	0.08	0.08	2
6	0.0843	0.0846	2	4	0.09	0.09	2
6	0.0848	0.0843	2	4	0.08	0.08	2
6	0.0844	0.0858	2	4	0.09	0.08	2
6	0.0841	0.0848	2	4	0.09	0.09	2
6	0.0867	0.0847	2	4	0.09	0.08	2
6	0.0853	0.085	2	4	0.09	0.09	2
				5	0.09	0.09	2
				5	0.09	0.09	2
				5	0.09	0.09	2
				5	0.09	0.09	2
				5	0.09	0.09	2
				5	0.09	0.09	2
				6	0.07	0.07	2
				6	0.07	0.07	2
				6	0.07	0.07	2
				6	0.07	0.07	2
				6	0.07	0.07	2
				6	0.07	0.07	2
				7	0.0883	0.089	2
				7	0.088	0.0888	2
				7	0.088	0.0889	2
				7	0.0886	0.0888	2
				7	0.088	0.0878	2
				7	0.0881	0.0884	2
				8	0.082	0.083	2
				8	0.083	0.085	2
				8	0.082	0.085	2
				8	0.087	0.087	2
				8	0.086	0.086	2
				8	0.086	0.086	2

Cr2O3 Lab	Duplicate 1	Duplicate 2	Cement	Method
2	0.027	0.027	1	1
2	0.028	0.027	1	1
2	0.027	0.027	1	1
2	0.027	0.027	1	1
2	0.027	0.027	1	1
2	0.027	0.027	1	1
2	0.012	0.012	2	1
2	0.011	0.011	2	1
2	0.011	0.011	2	1
2	0.011	0.011	2	1
2	0.011	0.011	2	1
2	0.011	0.01	2	1
3	0.0274	0.0274	1	1
3	0.0274	0.0264	1	1
3	0.0274	0.0264	1	1
3	0.0273	0.0273	1	1
3	0.0273	0.0273	1	1
3	0.0264	0.0273	1	1
3	0.0136	0.0146	2	1
3	0.0146	0.0136	2	1
3	0.0146	0.0136	2	1
3	0.0136	0.0136	2	1
3	0.0136	0.0136	2	1
3	0.0136	0.0136	2	1
6	0.024	0.025	1	1
6	0.026	0.025	1	1
6	0.024	0.025	1	1
6	0.024	0.024	1	1
6	0.024	0.025	1	1
6	0.024	0.024	1	1
6	0.01	0.01	2	1
6	0.013	0.012	2	1
6	0.01	0.01	2	1
6	0.01	0.01	2	1
6	0.01	0.009	2	1
6	0.012	0.012	2	1
7	0.0277	0.0286	1	1
7	0.0309	0.0299	1	1
7	0.0272	0.0291	1	1
7	0.0308	0.027	1	1
7	0.0296	0.0272	1	1
7	0.0287	0.0273	1	1
7	0.0126	0.0143	2	1
7	0.0115	0.011	2	1

143

7	0.0107	0.0133	2	1
7	0.0117	0.0131	2	1
7	0.0136	0.0136	2	1
7	0.0126	0.0131	2	1
8	0.02	0.02	1	1
8	0.02	0.02	1	1
8	0.02	0.02	1	1
8	0.02	0.02	1	1
8	0.02	0.02	1	1
8	0.02	0.02	1	1
8	0	0.01	2	1
8	0	0	2	1
8	0	0	2	1
8	0	0	2	1
8	0	0	2	1
8	0	0	2	1
1	0.03	0.03	1	2
1	0.03	0.031	1	2
1	0.029	0.03	1	2
1	0.029	0.03	1	2
1	0.031	0.031	1	2
1	0.032	0.031	1	2
1	0.013	0.013	2	2
1	0.014	0.013	2	2
1	0.013	0.013	2	2
1	0.013	0.013	2	2
1	0.015	0.013	2	2
1	0.013	0.012	2	2
4	0.016	0.015	1	2
4	0.016	0.014	1	2
4	0.015	0.015	1	2
4	0.016	0.016	1	2
4	0.014	0.015	1	2
4	0.015	0.015	1	2
5	0.028	0.029	1	2
5	0.028	0.028	1	2
5	0.027	0.027	1	2
5	0.028	0.028	1	2
5	0.027	0.027	1	2
5	0.027	0.027	1	2
5	0.012	0.012	2	2
5	0.011	0.012	2	2
5	0.011	0.011	2	2
5	0.012	0.011	2	2
5	0.012	0.012	2	2
5	0.011	0.011	2	2

Acknowledgements

We wish to recognize LeRoy Jacobs, who provided the enthusiasm and his talents to the ASTM C1.23 task group in coordinating the inter-laboratory study. The time, efforts of all of the participants in this trial program and Robin Haupt of the Cement and Concrete Reference laboratory for sample distribution and collecting data are acknoledged. The comments and suggestions of reviewers Clarissa Ferraris, Kenneth Snyder and Don Broton are gratefully acknowledged. This project was supported by the Early-Age Performance of Concrete project within the Sustainable Engineering Materials program at NIST.

References

[1] Data Considered by Committee C-1 of the American Society for Testing Materials in Preparing the Standard Specifications and Tests for Portland Cement (serial Designation C 9-17), Committee C-1, American Society for Testing Materials, Philadelphia, PA, July 1919.

[2] C114-11b, "Standard Test Method for Chemical Analysis of Hydraulic Cement," Annual Book of ASTM Standards, Vol. 4.01, ASTM International, West Conshohocken, PA.

[3] J.A. Forrester, T.P. Lees, and A.E. Moore, "The Precision of Standard Cement Analysis nd its Effect on the Calculated Compound Composition," S.C.I. Monograph No. 18, The Analysis of Calcareous Materials, 447-451 (1964)

[4] H.G. Midgley, "Compound calculation in the phases in Portland cement clinker," Cement Technology, Vol. 1, No. 3, 1970, pp. 1-5.

[5] A.M. Harrisson, H.F.W. Taylor and N.B. Winter, "Electron-Optical Analysis of the Phases in a Portland Cement Clinker, with some Observations on the Calculation of Quantitative Phase Composition", Cement and Concrete Research, Vol. 15, pp. 775-780, 1985.

[6] L.P. Aldridge and R.P. Eardley, "Effects of analytical errors on the Bogue calculation of compound composition," Cement Technology, 4, 1973, pp. 177-182.

[7] L.P. Aldridge, "Errors in the analysis of cement," Cement Technology, 7, 1976, pp. 8-11

[8] P.E. Stutzman, and D.S. Lane, "Effects of Analytical Precision on Bogue Calculations of Potential Portland Cement Composition," Journal of ASTM International, Vol. 7, No. 6, paper ID JAI102697

[9] EN 196-2, Methods of Testing Cement – Part 2: Chemical analysis of cement,

[10] NCHRP 139, "Precision Estimates for AASHTO Test Method T 105, Determined Using CCRL Proficiency Sample Data". NCHRP Project 09-26A, June 2009, http://onlinepubs.trb.org/onlinepubs/nchrp/nchrp_w139.pdf

[11] Report of Working Committee on Methods of Chemical Analysis, June 12, 1946, W.C. Hanna, Chairman

[12] R. Haupt, personal communication

[13] ASTM E 691 "Standard Practice for Conducting an Interlaboratory Study to Determine the Precision of a Test Method," Annual Book of ASTM Standards, Vol. 14.02, ASTM International, West Conshohocken, PA.

[14] ASTM E177 "Standard Practice for Use of the Terms Precision and Bias in ASTM Test Methods," Annual Book of ASTM Standards, Vol. 14.02, ASTM International, West Conshohocken, PA.

[15] G.J. Hahn and W.Q. Meeker, *Statistical Intervals: A Guide for Practitioners*, John Wiley and Sons, New York, 1991.

www.ingramcontent.com/pod-product-compliance
Lightning Source LLC
Chambersburg PA
CBHW080251180526
45167CB00006B/2492